Motorbooks International

POWERPRO S[ ]

HOW TO BUILD & MODIFY
CHEVROLET
SMALL-BLOCK V-8
PISTONS, RODS & CRANKSHAFTS

David Vizard

To those valuable friends who
helped me through
some very difficult times

First published in 1992 by Motorbooks International Publishers & Wholesalers, PO Box 2, 729 Prospect Avenue, Osceola, WI 54020 USA

Motorbooks International books are also available at discounts in bulk quantity for industrial or sales-promotional use. For details write to Special Sales Manager at the Publisher's address

Library of Congress Cataloging-in-Publication Data
    Vizard, David.
      How to build & modify Chevrolet small-block V-8 pistons, rods, and crankshafts / David Vizard.
          p.   cm.—(Motorbooks International powerpro series)
    Includes index.
    ISBN 0-87938-579-0
    1. Chevrolet automobile—Motors—Modification. 2. Chevrolet automobile—Motors—Pistons and piston rings. 3. Chevrolet automobile—Motors—Crankshafts. I. Title. II. Title: How to build and modify Chevrolet small-block V-8 pistons, rods, and crankshafts. III. Series.
    TL215.C48V59   1992
    629.25′04—dc20   91-41716

On the front cover: The Chevrolet small-block is a POWERPRO-prepared 010 heavy-duty truck block fitted with Summers Brothers angle-bolt, four-bolt bearing caps. The crankshaft is from Callies Performance Products and is ion-nitrided with a five-point oiling system for high rpm and long endurance. The piston is from Wiseco for an 18 degree Chevrolet Bow-Tie cylinder head. The connecting rod is for Winston Cup racing from Callies Performance Products. *Tom Rizzo*

Printed and bound in the United States of America.

About the Author

Author David Vizard is an experienced engine-builder, road-race and drag-race engine tuner, and engineer, having built high-performance engines for all types of cars from Camaros to dragsters to Chevrolet engines for a Lola T70 sports-prototype racer.

His workshop in California is equipped with a 1000hp computer-controlled Superflow dyno, flow bench, and full machining and welding facilities, allowing him personally to experiment with and test the products and procedures covered in this book. Unless otherwise noted, all test results, tables, drawings and photographs in this book are by David Vizard.

The author has written more than 3,000 magazine articles and 21 books, including **How to Build & Modify Chevrolet Small-Block V-8 Cylinder Heads**, also published in Motorbooks International's POWERPRO Series.

Contents

Acknowledgments and Guest Editors

I would like to thank all of the people who so willingly helped me make this book possible. Thanks to Jeff Roper, Ian and Ken at Cosworth, Howard Stewart, Jack Evans, Denny Wycoff, John Callies, Dan Swain, Gary Thompson, Linda Erath, Henry Velasco, Darryl Beuhl, Allan Nimmo, Jerry Weaver, Paul Smith, Mike Parry, Jerry Goodale, Kevin McFadden, Carl Schattilly and Bob Benton.

I would also like to thank my guest editors for their assistance.

John Erb

John Erb is involved full time with the science of engine building. His earlier design and development work with supercharged engines was especially helpful with piston design parameters, and he is the designer of the Keith Black Signature Series of pistons, as well as the chief engineer of Silvolite United Engine & Machine Co.

John is a pilot and an Experimental Aircraft Association member, and holds the patent on an Avionic lock as well as being the designer of high-speed book manufacturing equipment, oil-free cannery food-processing equipment and a new concept chip circuit test station. In addition to this, he has a maintenance-free solar-heated home, and it's this varied background that gives him the ability to take a fresh, new look at the everchanging engine performance challenge.

Moe Mills

Moe Mills first became associated in the speed equipment industry in 1966 when he took a job with Crank Shaft Co., builders of the original welded strokers, in Los Angeles. His engineering talent led to a rapid series of promotions over the next five years, and he became general manager.

After later holding positions at Scat Enterprises and one of the larger racing piston manufacturers, he and Ken Roble established Ross Racing Pistons in 1979.

His hobbies include running his top fuel dragster and flying his twin-engined Cessna.

Billy Howell

A 1961 engineering graduate of the University of Wyoming, Bill Howell joined Chevrolet Engineering as a laboratory test engineer. He spent the first six years of his Chevy career directly involved in engine development and durability testing. During that time he was the sole engineer on the Mark II 427 and 396 Mystery engine, debuted at the 1963 Daytona 500, and Mark IV Hi-Perf engines from 1963-1967 in all variations.

He moved to Vince Piggin's group in 1967. This development was the back door through which all of Chevrolet's racing programs were conducted. Continuing as a development engineer, he monitored engine test programs with Smokey Yunick, McLaren Racing, Katech, Ryan Falconer Racing Engines, Dennis Fisher Engines, Traco Engineering, Roger Penske, Pro-Motor Engineering and several others that had Chevy programs. He was also the originator of most of the text and parts list in the Chevrolet Power Catalog.

Retiring in 1987, he and a partner have formed a company to develop and sell wiring harnesses to fit Chevy fuel-injected production engines into other vehicles and boats. He also does engine development consulting work.

Allan Lockheed

Allan Lockheed is author and distributor of **The Engine Expert** program for piston engine design. Many top racing engine builders use the program, which is gaining widespread recognition in the performance industry.

Allan's background includes motorcycles, 427 and 351 Fords, Shelbys, oval-track stock and modified Chevys, and most recently the 2.2 Chrysler turbo. His degree includes computer science and mathematics, and he established technical computing departments for Exxon Minerals and Union Pacific Mining. He is looking forward to extending his work with the performance industry to include aircraft.

Allan has a life-long fascination with high-performance vehicles and people—inherited from his father, founder in 1927 of the Lockheed Aircraft Company, and his uncle Malcolm, inventor of hydraulic brakes.

Two of Allan's favorite engine projects have been a vintage racing Lotus twin-cam Elan and a stroker 452 ci Ford tunnel-port for a street Pantera.

Dick Howell

Dick Howell is by occupation an engineer working on ballistic missile systems for TRW in San Bernardino, California. He is also president and chief engineer of the Howell Engineering Company. This company deals almost exclusively in the mathematical analysis of engine related subjects.

His current publications include a book, **Automotive Engine Piston, Connecting Rod, and Crank Dynamics,** and an engineering monograph entitled, "General, Exact and Analytic Instantaneous Torque Equation for Any Single or Multi-cylinder 4-Cycle Even Firing Reciprocating Engine."

Amongst other qualifications, Dick has a BSEE and BSME from West Coast University, and he has done post-graduate work at CalPoly, in mechanical engineering.

Jim Walther

Jim Walther began racing professionally in 1960 when he came out of the Army. He raced midgets for three years, but then had his first ride in a top fuel car, and from that moment on was hooked on power and speed. He stopped oval-track racing, and built his first top fuel car in 1963, and was successful, winning the NHRA Top Fuel World Champion title in 1972. He has run with every major drag race organization through the years.

Jim started with Wiseco in 1983, where he is now the automotive sales manager.

Engine Block Selection

1

No matter how you look at it, the block is the foundation of a high-performance engine. For whatever application, it will pay to think about what is needed in terms of target horsepower *before* choosing a particular block. Thus, rule number one is: *Do not be tempted to use a block just because you happen to have it.* Many people build weird combinations that are less than optimal for whatever requirement they have; when questioned as to why such a combination was chosen they answer that, "I just happened to have that block." Remember, production blocks are cheap. You can go to the wrecking yard and pick up a short motor for less than $100 for most applications.

Chevrolet Small-Block Family Tree

The small-block was introduced in 1955 with 265ci. Unless you are restoring a car and want to be exact in every detail, the early 265 block is not one to use since it has no oil filter. During 1956 this was rectified and the engine continued in production until 1957.

In 1957, the engine size was stepped up to 283ci. This engine, with its 3.875in bore and 3.000in stroke, continued production until 1967. Up to about 1958 the small-block Chevrolet used a rope-type rear main seal, but from this point to the 1986 model the now-familiar "split" neoprene seal was used.

In the 1962 model year the Corvette came out with 327ci. This was the first of the 4.000in bore motors, and set the stage for what was to become the most common bore size among small-blocks. Equally noteworthy is the fact that the stroke was increased to 3.250in. This necessitated larger counterweights and as a result, some substantial internal modifications had to be done to the 283 design to accommodate a crankshaft with larger counterweights.

In the 1968 model year when the 350 was introduced, blocks moved into the big-journal crankshaft era and the 327 was produced both in small- and big-journal size. This means if you have a 327 big-journal crank, you can make a 327 engine out of a 350. Later 327s from 1968-1969, when the engine was dropped, had bigger main-bearing journals than earlier engines.

In 1968, the 307ci engine was introduced as a kind of economy engine. It had the 3.875in bore of the 283 and used a cast crank with the 3.250in stroke of the 327 and was produced up to 1973.

From 1967 to 1969, the 302 was produced, although it was probably developed just so Chevrolet had an engine size eligible for Trans-Am racing. In 1967, it was produced with the small–journal crank and in 1968-1969, with the big–journal crank.

The year 1967 also saw the introduction of the 350ci engine. This now-classic configuration utilized a 4.000in bore with a 3.480in stroke, and has probably become the most common engine size. Along with the introduction of the 350 came the standardization on the big main journals.

The next major milestone in terms of performance was the introduction in 1970 of the 400ci block. This deviated from the standard format in many respects. First, it used a nominal 4.125in bore, as opposed to the 4.000in bore that had become the norm. To achieve the 4.125in bore, some internal casting core changes were necessary. The most obvious external change was the use of three freeze plugs in the side of the block. To

The basis for any high-performance engine must be a sound bottom end. This 450hp street 350 that I built had selected stock components suitably reworked for the required reliability.

accommodate this larger bore size, the bores were siamesed; there was no water between the cylinder bores because the barrels joined in the water jacket. In contrast, all other blocks have water completely surrounding each cylinder bore.

Along with this change in the bore size, the 400 motor was equipped with a 3.750in stroke cast crankshaft. To get this bore-stroke combination within the confines of the standard 9.025in crankshaft center to block deck height it was necessary to shorten rod length if the stock piston ring package was to be retained. The standard 5.70in rod was then shortened to 5.56in.

In 1975, the 262ci engine was introduced with a 3.671in bore and a 3.100in stroke. It was intended as a low–output economy engine but obviously did not prove popular as it was produced for only one year.

In 1976, the 305ci engine was introduced, which must be the most plentiful engine in existence, next to the 350. Built largely for emission and mileage reasons, this engine sports a 3.763in bore with the crankshaft stroke of the 350 at 3.480in. As of 1992, the 305 is still being produced and looks as if it will continue in production for some time.

In late 1979, a 267ci engine was introduced. This had a 3.500in bore along with the 3.480in stroke of the 350. Like the earlier 262, it was intended as an economy engine, both in terms of cost and fuel consumption. Again like the 262, it did not prove popular and was produced only through model year 1982.

As of 1983, only two displacements of small-block were produced—the 305 and 350. In 1986, a block design change was made concentrating on the rear main oil seal. Instead of having a split oil seal, the rear main bearing now used a full 360deg. one-piece seal. Using this type of block usually requires using the relevant crank and flywheel, as a crankshaft design change was also made. If you have an early crank, Chevrolet has a special seal adaptor kit to convert 1986 and the late block to accept pre 1986 cranks.

A hydraulic-roller version of this new block was also introduced at about the same time. The roller-follower assembly is not interchangeable with later blocks. Unlike aftermarket rollers that are prevented from rotating by linking them in pairs, the General Motors roller setup has special slotted plates that locate lifters and keep the rollers aligned with the cam lobe.

Main Bearings

When it comes to blocks for a high-performance engine, the number-one choice is a four-bolt main-bearing block over the more common two-bolt. Chances are, unless you pick it up from someone who doesn't know what they're selling, the cost of a four-bolt block will be significantly more than a two-bolt. For high-performance applications the four-bolts are certainly better, but there is a definite point below which a four-bolt setup is needless overkill.

The basic two-bolt block for a drag-race application can successfully stand as much as 500hp; for high-performance on the street it appears to be satisfactory at power levels up to 400hp. For a circle-track application where induction restrictions such as the use of a two-barrel carburetor limit power to no more than about 375hp, the two-bolt block will get the job done. If the limitations on a circle-track motor call for carburetor and exhaust restrictions that limit power to 300hp, then the two-bolt block is every bit as reliable as the four-bolt.

Casting Material

Equally important in terms of wear is the type of cast-iron alloy the blocks are made of. The most desirable are the high-nickel blocks as the high-nickel content promotes surface hardening of the bores, which greatly extends bore life. The question is, how do we identify these blocks? This actually proves easier than you would expect. In the area normally under the cam timing gear, there are numbers that give away the show. What you're looking for is 10s and 20s. A 10 over a 20 indicates a block produced from a cast-iron alloy having 1 percent tin and 2 percent nickel.

The tin helps in the pouring of the block and makes the cast iron more ductile, that is, less prone to cracking. Because blocks containing tin tend to cast more easily, they tend to have less problem with hot spots caused by porous metal.

The nickel content of the iron alloy used in the 10/20 block does two things: it refines grain structure of the cast iron, increasing its strength and work-hardening characteristic. As the rings rub up and down the bore it tends to harden the surface, resulting in considerably extended bore life.

If the block has only one number, either a 10 or 20, this means it contains no deliberately added tin and has either 1 or 2 percent added nickel. If no number is present, then no additional nickel has been added to the cast iron. I've heard some 307ci blocks were cast with this basic material; if this is true, it would account for the higher bore wear rates of some of these engines.

Choosing an Engine Size

Now the question is, What displacement to choose? The answer obviously depends on the intended application.

The most popular usage for a small-block Chevy is as a performance truck or car street motor. The first thing to remember is, there is no substitute for cubic inches unless it carries with it some design or cost penalty. For instance, it is pointless putting money into a 302ci engine if for a small amount more a 350 can be had. Keep in mind, a good street motor needs a power curve that starts as low as possible and anytime a displacement is reduced, the low-end power decreases significantly.

Having just set a ground rule for cubic inches, you might feel safe in assuming the 400ci engine would be the one to go for but this is not necessarily the case. If a tight budget is a criteria, then a 400 is *not* a good choice. Obtaining and prepping the 400 block costs a lot more money than a 350. Due to its siamesed bores a 400 needs extensive work to make it reliable in producing adequate cylinder-to-cylinder cooling at high outputs. Thus for simple preparation and near troublefree results, a 350 is the way to go.

Assuming that finances are not critical, the choice of engine must be based solely on how successful it's likely to be in the application. For instance, let's say a drag-race engine is to be built and the class calls for a certain number of pounds per cubic inch. It is important to keep in mind that any cylinder heads likely to be used on a small-block Chevy do not have anything like the airflow we would like to see for *real* race engine outputs. So when putting together an engine that must compete on a pounds per cubic inch basis, make sure that the bottom end of the engine does not significantly outpace cylinder-head flow capability.

For drag-racing applications, displacements around 300ci appear to net the best horsepower per cubic inch. To achieve good results at this displacement, make sure that the type of cylinder heads used and modifications employed can result in a small enough combustion chamber to achieve the high compression necessary—for instance, around 14.0:1 to 16.0:1—without compromising the chamber with a high piston dome.

If a circle-track or road-racing engine is required, then horsepower for an extended period of time is more important and inevitably means operating the engine at less rpm. At lower rpm levels, a displacement of around 330ci nets the best horsepower per cubic inch.

Moving into off-road racing, yet another set of criteria must be dealt with. Here the engine must have good power at relatively low rpm, which helps to avoid engine bog when tackling soft, steep or heavy-going terrain. For such use, the more cubic inches the better—but there is always a price to pay. Most successful off-road motors are at least 406ci, but the most common *race-winning* size is around 430.

Putting this many cubic inches into an engine that has only a 9.025in crankshaft center to deck height, however, entails the use of a rod-to-stroke ratio that is on the small side. This causes a lot of rod angularity and leads to more friction between the piston and cylinder wall. The net result is that these motors tend to scuff and wear out bores far quicker than their shorter-stroke counterparts.

If you choose a bracket-race motor, then displacement will depend on the weight of the vehicle and the time bracket you intend to run in. The main ingredient for bracket racing is consistency and the most consistent motors tend to be those that are not maxed out. Thus, when a bracket-race engine is desired you should not be shooting for the 2.1hp per cubic inch that the 292ci Buick-headed econo-rail racers have.

Finding the Right Engine

Unless you already have the block of your choice, the first step should be boning up on block identification so you can locate one. The easiest way to

do this is by casting number; use the block number chart to identify what you are looking at.

Be aware that when you are looking at block numbers, it will be necessary to choose a block that has the relevant external fittings for your vehicle's needs. For instance, a threaded hole for the clutch linkage will be needed for a manual transmission and, if the block is to go in a Chevy II, a compatible oil filter system will be necessary, as this block employs a deeper-set filter for chassis rail clearance. Other variations such as dipstick location also need to be taken into account.

The most obvious place to get a working core is the wrecking yard, but it's not the only source. I've managed to procure several late-model engines from a mining company that has a large truck fleet as well as regular passenger cars for service vehicles. These accumulate high mileage quickly under arduous conditions, so the turnover is rapid. I don't deal with the mining company directly but I do deal with their appointed engine reconditioner; by telling them what I'm looking for, they let me know when a suitable core becomes available.

Engine Disassembly and Testing

Once a block has been procured, make sure it's drained of oil, then drag

it off to the nearest power car wash and spray it with Gunk engine cleaner. Brush it in and blast it with high-pressure water. Doing this job first means less mess in your workshop.

Once the engine is back in your shop, mount it on an engine stand, *not* on a bench. These days, engine stands are so inexpensive that there is no excuse for not having one. The cut-price warehouses are usually a good, inexpensive source for engine stands.

Begin by stripping the core motor to the bare block, numbering rods and caps as you go so they can be returned to the identical location on rebuild, and noting the condition of the parts as these can be good early-warning problem indicators. Pay special attention to the piston-skirt wear pattern: if it is uneven on the thrust faces, there is a fair chance that connecting rod is bent. A bent rod increases bore friction and cuts ring seal; a bend producing 0.005in error across the width of a close-fitting piston can sap 20hp from the engine. If a rod does show signs of being bent, set it aside for possible correction.

Once the block is stripped bare it should be checked for bore wear and cracks. Skip this last point and you could pay dearly. A particular area to check is around the bottom of the bores. If the engine was once rebuilt by someone not familiar with the small-block, they could have tried to

All the oil galleys will need a thorough cleanout. To do that it's necessary to remove the tapered, threaded plugs from various parts of the block. For the most part, the easiest way to get these out is to heat up the plug with a welding torch until it just starts to glow red-hot, then let it cool.

Chevrolet Small-Block Casting Identification

Casting Number	CID	Years	Vehicles
360851	262	1974-1976	Monza
3703524*	265	1955	Passenger cars
3720991	265	1956-1957	Trucks and passenger cars
14010280			Passenger
14016376	267	1979-1982	cars
471511			
3731548	283	1957	
3556519			Trucks and
3737739	283	1958-1961	passenger
3849852			cars
3789935			Trucks and
3849852	283	1962-1964	passenger
3864812			cars
3849852			Trucks
3849935	283	1965-1967	and
3896944			passenger
393288			cars
	302	1967	302 Camaro
	302	1968-1969	302 Camaro
389257***			
14010201			Passenger cars and
14016381	305	1980-1984	light trucks
14010202			
14010203			
460776			Passenger cars and
460777	305	1978-1979	light trucks
460778			
361979			
3914653			Trucks and
3914636	307	1968-1973	passenger
3932373			cars
3970020			
3959512	327	1962-1963	
3782870			Trucks
3789817	327	1962-1964	and
3794460			passenger
3852174			cars
3858180			
3892657			Trucks
3782870	327	1964-1967	and
3903352			passenger
3789817			cars
3858174			
3892657			
3791362	327	1965-1967	Chevy II
3970041			Corvette
3814660			Camaro and other
3970010	327	1968-1969	high-performance
3914678			applications
3932386			
3955618			
3855961			Passenger
3932388			cars
3958618	350	1968-1976	(two-bolt mains)
3970014			
6259425			
3956618			Truck and high-
3970010**	350	1968-1979	performance applications
3932386			(four-bolt mains)
14016379	350	1978-1979	Passenger cars and
366245	350	1978-1979	light trucks (dip stick in pan)
140029	350	1980-1984	Passenger cars
14010207	350		
3951511	400	1970-1973	Heavy-duty trucks and passenger cars (four-bolt mains)
3951509	400	1974-1976	(two-bolt mains)
3030817			

*First six-months' production used mechanical cams instead of hydraulic.

**Sometimes machined for two-bolt mains.

***Small-journal 302ci engine uses 327 block. Large-journal 302ci engine uses 350 block.

drive out the connecting rods without realizing the rod can hang up on the bottom of the bore, possibly cracking it. In most cases this does not compromise the block except making it look ugly, but occasionally such guerilla tactics can cause a crack in the water jacket. Also, a connecting rod breakage can compromise the block, so look for damage in these areas, assess it and make a realistic decision as to whether the block can be used.

Since the block is such an important part of any engine build program, it's necessary to ensure that it is up to snuff, so be on the safe side. It is pointless to put money and time into a block that is even vaguely suspect.

To assess the viability of using a particular block first measure the bores. If the bore size is already at 4.030in with a 0.030in overbore, the engine should be rejected. I don't like to use 4.000in or 4.125in bore blocks for high-performance applications past an 0.030in overbore. This is especially so for a 400 block. The reason is that bore flex begins to have a noticeable effect over both of these sizes, especially if there is any core shift involved. The block can be sonic tested—a procedure recommended for race engines—but it costs money, sometimes more for a street motor than the cost of getting another block.

The best block is one that has never been bored as this will allow you to bore to 0.020 or 0.030in oversize.

After bore size and the lower block crack situation have been investigated, turn your attention to the main-bearing saddles. What you're looking for here is a good fit of the main bearings in their location slot.

If the caps are loose in the location slot, then obviously repair is necessary. At this point you must decide whether the block is good enough to fix or if another block should be selected. If the cap is loose in the block —and some come this way from the factory—some engine machine shops center-pop the edge alongside of the cap. This spreads some metal back into the groove, thereby tightening the cap.

I don't like this technique, but sometimes there's no option. If loads incurred by normal use or a main bearing failure were sufficient to pull the cap in at the sides in the first place, then center-popping the side of

the groove is not going to provide a press fit for long. A better, but obviously more time-consuming and costly way is to grind a little off the side of the offending caps, weld them up with some hard material such as stainless steel, then grind the edge of the cap on one side until the bore just lines up with the mating half of the main-bearing bore in the block. Once one side has been aligned accurately for line honing to take out any *minor* remaining error, then the other side of the cap can be cut until it fits snugly in the block. There aren't many shops that perform this type of work, so don't count on it being done by your present machine shop. Some telephoning around may be required.

Next, check to see that the threads for the main-bearing cap bolts are in good order. If the threads are stripped, then it's possible to successfully heli-coil the block, although if the block takes more than one heli-coil per cap, think about discarding it because although the heli-coil is strong, the area around it will have weakened some.

Finally, look hard and long at the area below the saddle in the block to establish whether or not any cracks exist below the cap area from overtightening or stressing due to use. Look for cracks originating from the outside bolt holes of the three middle main caps on a four-bolt block.

If the block passes all the tests so

far, then remove it from the stand to facilitate the removal of the oil galley plugs. The ones that are particularly difficult are the square-socket screw-in plugs that are installed in the back of the block and around the oil filter. You may also have a problem removing the coolant drain plugs from the center side of the block toward the pan rail.

Two different techniques can be used to remove stubborn plugs. They can either be drilled out, or pre-loosened with an oxyacetylene torch. I prefer the torch technique, since it is much quicker and threads are less likely to be damaged as could be the case with misdirected drilling.

The torch technique involves using a narrow flame to heat the plug as hot as possible to red heat. The tight fit of the block on the thread then tends to squeeze the thread down slightly so when it cools off the pressure on the thread is reduced. This speedy method seems to work ninety-nine times out of 100, and in half the cases the first application of heat allows the plug to be broken free.

Once all the screw-in plugs are out of the block, strip out the press-in and freeze plugs and punch the head dowels through the deck into the water jacket. When dealing with the front oil galley plugs, be sure to deburr the edge of each hole first

After the heat treatment the plug will usually break free relatively easily. If it fails *to do so, then you'll have to resort to drilling it out.*

because these plugs are normally pressed in, then deform the edge of the hole with a cold chisel for 100 percent retention.

After removing the cam bearings you should be down to a bare block, and if it has passed all the basic tests it should be functional—at least for a high-output street motor.

Block Cleaning

What the block now needs is a thorough cleaning before starting any machining operations. The normal technique for this is to boil out the block in a hot tank. If the chemicals are fresh, this rids the block of debris and should scour the oil passages down to bare metal.

A preferable alternative is a jet clean tank. However, disposing of cleaning chemicals is becoming more complicated as the Environmental Protection Agency, or EPA, rightly insists that such chemicals are dealt with in an ecologically safe manner. This has prompted many machine shops to look for alternative cleaning systems that do not involve chemicals and the resultant chemical waste. Cleaning ovens, for example, are becoming a popular alternative.

Cleaning ovens heat the block to approximately 500deg. Fahrenheit. The deposits are then baked to the point where they are dried up and ready to fall off the block. After heating for 20-25 minutes, the block is then put into a steel shot cabinet, which rotates the block while steel balls are hurled at it. The velocity of the balls is controlled so that they don't damage the metal, but they do hit the block hard enough to remove any residual deposits. When finished, the block will look brand new, with all rust and scale totally removed.

If your cleaning operation does not get the block down to the cast iron, then a wire-brush scrubbing will be necessary. A wire brush that is just a bit larger than the freeze plug holes is best.

Once the block is clean, there is plenty of detailing to be done. For this, an electric die grinder and tungsten

The typical ovenbaking cleaning technique involves mounting the block in this cradle. The block is then lowered into the furnace. The baking treatment involves heating the block to some 550deg. F., which bakes all the residues to a cinder.

After baking the block goes into a shot-blasting cabinet, which removes any remaining residue from the baking process down to the bare metal of the block.

To aid oil drainback, the back end of the block can be ground as seen here.

carbide cutter get the job done. If you are not already suitably equipped, a Makita grinder is a good choice because of its low cost.

Final Block Preparation

Some of the following procedures are not imperative, but they do help improve the appearance of the block. The first thing I do to get a clean, stripped block is to de-flash the outside of the block. There are two ribs on the front corners of the blocks where the various patterns pull away during the casting procedure; another is above the oil filter mounting. Grinding these off and blending them in tidies the block considerably.

Next, turn your attention to the block valley. Here I do quite a bit of work on the block to clean up the areas where sand inclusions predominantly occur, and to aid oil drainback. Check down at the back of the block where the distributor bore is located; there is a square-shaped hole there where a split line occurs in the casting pattern, which often hides a lot of dirt. Getting in there with a carbide cutter and taking it down to bare cast iron will rid the casting of possible sand inclusions that may break loose.

At the front of the block, rework the two round holes that connect the area under the cam timing cover with the valley. Drop the lower edge of the holes and radius them off so any oil that may have moved up to the front of the engine during braking drains back via the cam gear. Also, the pads at each corner of the valley must be angled down, reshaped, matched to the drain-back area in the cylinder head and polished.

To help establish the area to be reworked, place the head on the block and mark the oil drain-back hole at the ends of the cylinder head through to the block. Then grind and radius the hole for a smooth transition for oil flow back to the pan.

By this point you should have considered what type of rockers the engine is to be run with. If you plan to use needle-bearing roller rockers along with sustained high rpm, then some precautionary measures are in order. It is not uncommon for needle-bearing rockers to break up and dump the needle rollers into the oiling system. To avoid making a bad situation worse it is worthwhile to install a screen in the valley oil return holes to catch any debris en route to the oil pump. If regular-type rockers are to be used along with moderate valve springs, a screen is not essential. Heavy-duty valve springs may sometimes break a tail, so protecting the pump from spring debris is important.

Oil Galley Preparation

The oil pump pressure and flow can be better communicated to the vital areas, such as bearings, by porting the galleys connecting the oil pump to the bearings. The point to begin this operation is at the recess in the rear main-bearing cap where oil is received from the pump. Oil discharged from the pump must flow around the pump fixing bolt, then down the angle drilling on the side of the cap and into the block.

Sometimes the cavity combined in the pump and cap around the bolt is shallow and restricts oil at higher rpm. To resolve this problem, match the cavities of the pump and cap (for details on reworking the oil pump, see chapter 11). Blend the oil drilling in the side of the cap where it meets the cavity to allow easier oil flow. Next, pull the cap from the block and check the alignment of the cap hole with its opposite number in the block. With a well-used block any misalignment is usually easy to see because oil de-

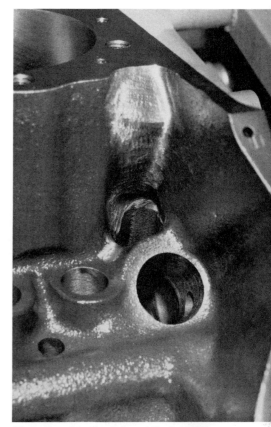

Work down into the holes of the base of the valley as far as possible because this is an area where sand inclusions are often to be found.

Compare these two rear mains caps. With the one on the right you can see that the recess around the pump retaining bolt has been cleaned up, made deeper, blended into the oil hole going down into the cap and matched to the pump. All this is in an effort to improve oil flow.

posits discolor the mismatched areas. Using this as a guide, match the holes in the block and cap.

Just after the oil enters the block it must do a near right-angle turn. Using a suitably sized ball-ended carbide cutter, go through the hole just opened up and through the access drilling in the side of the rear main housing and radius off the edge of the hole so the oil can more easily make the turn into the horizontal galley.

From here the oil travels along this drilling and intersects at a shallow angle with the drilling from the filter housing. Often these drillings wander somewhat and a ball-shaped carbide cutter on the end of a long shank proves a useful tool for aligning the two drillings and smoothing the transition from one to the other at the intersection point.

Flare the point where the drilling from the oil pump enters the filter cavity to ease the flow of oil out and around the top of the filter. From here the oil will flow down from the holes in the filter case, through the filter and up through the middle. Any edges at the filter-mounting center cavity where oil reenters the block from the filter should be radiused off. Also radius the edges of the horizontal drilling that leads the oil away from the filter cavity.

The next step is to check the alignment of the holes in the main-bearing saddles with the bearing shells. Most bearing shells are reasonably consistent as far as the oil hole is concerned, so use an old bearing shell, and grind the exit point of the hole in the block so it matches the bearing shell. This ensures that no unnecessary oil flow restriction takes place.

For the moment let's assume our motor will be a stock stroke unit, and rod-to-block clearance will not be a problem. If this is the case, we're just about done as far as grinding is concerned. If we intend to install aluminum rods, a stroker crank or both, some block grinding for clearance will be required.

The rear main cap on the left has had the oil transfer hole to the block reworked to match the block, whereas the cap on the right has not. Also, just discernible, is the step where the angle drilling meets the vertical drilling into the cap. This needs to be blended out to improve flow.

Tapping Threaded Block Holes

A lengthy tapping and thread-cleaning exercise is now called for. The first step requires a ¼in pipe tap; a second or bottom-cut tap will also be required. Using the ¼in tap, tap the three oil galley holes at the front face of the block above the cam gear bearing bore. These holes are, strictly speaking, oversized for tapping with this size of tap, but sufficient thread will be cut to allow a socket-drive taper plug to be installed. Screw-in plugs simplify galley cleaning at future teardowns. Tap the galley holes only to the step formed about ½in from the front block face. When selecting plugs, be sure they are short enough so as not to restrict the intersecting oil holes.

If a roller cam is used, the oil drain-back holes down the center of the valley can also be tapped to accept plugs or stand-off tubes that allow

This view, looking into the oil filter housing, shows how entrance and exit points for the oil are reworked to improve flow.

blowby up but not oil down; Y4 pipe nipples work well. The reason these holes can be plugged is that a roller cam requires minimal lubrication at the roller-cam lobe interface; more than enough oil will be thrown up to lubricate the followers by the rotating parts in the crankcase. Plugging these holes also helps keep the drain-back oil from being caught up in the crank.

From here on anything to do with threads in the basic block preparation involves cleaning, and the best way to do this is by retapping all the holes. When it comes to the threads for the cylinder-head bolts, these should be given special attention. Even if inexpensive taps are used elsewhere to clean the threads, at this point use a high-quality tap in mint condition as cheap or blunt taps may cut a rough or oversized thread. This can lead to reduced clamping pressures, thread failure or both, either of which leads to head gasket failure. By using a good-quality tap you get the best chance of reproducing a thread that is of the high standard the job requires.

Next, load a chamfering tool into an electric drill and chamfer *every hole* in the block. With threaded holes the aim is to remove the top thread so bolt tightening doesn't pull the top thread and prevent seating of whatever is being clamped down.

The most important threaded holes to chamfer are the head bolt holes. If the block is to be decked of any significant amount, then the holes will need to be chamfered an appropriate amount deeper so that sufficient chamfer remains after the machining of the block faces.

When all chamfering is completed, go around the block with a flat or half-round needle file as appropriate and *lightly* chamfer all the sharp edges left from the machining operations—the most important being those on the main-bearing caps and saddles. A 0.005–0.010in chamfer is all that is needed to get the job done.

The block is now ready for any machining that must be done, so the main-cap mounting surfaces should be cleaned and the main caps installed in their numbered order and the bolts appropriately torqued down. The block is now ready to ship off to a machine shop.

Machining the Block

Before any machining can be done, consider your budget for the project.

Unless you have an unlimited budget, you need to establish some priorities. For starters, the block will definitely need to be rebored and honed. Additionally, if the positive effect of the cylinder-head quench area is to be maximized, then the block is going to need decking so the pistons make the desired close approach to the cylinder-head face. Main-bearing alignment might also need honing for accuracy. The point is that all these machining operations cost money. Boring and honing the block is obviously mandatory and, depending on the pistons chosen, the decking and milling operation may or may not be required.

Of primary concern, though, are the main bearings. Most high-performance engine builders have the main bearings line-honed as a matter of course. As they come from the factory, however, most blocks are good and except for the odd one, line-honing the mains can be viewed as gilding the lily, at least for a street motor.

Before considering line-honing if you're on a tight budget, check to see whether this operation is required. The simplest way is to use a finished crank and new bearings for a spin test. Assemble the bearings and crank into the block and lubricate all surfaces with a lightweight oil. Begin-

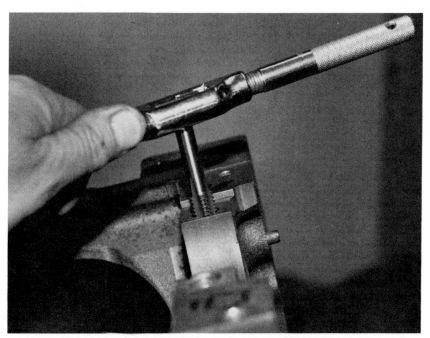

To develop the required clamping forces, main bearing and cylinder head threads must be clean. The best tool for the job here is a new, high-quality tap.

Tap Sizes and Tapping Locations

Sizes	Location
1/4x20	Pan rail and timing cover
5/16x18	Front and rear of pan rail
3/8x16	Front of block
7/16x14	Head bolts and mains bolts
3/8 Pipe	Front and rear of main and lifter galleys water drain both sides at bottom of block, top rear galley and filter housing
1/8 Pipe	Front and rear oil pressure takeoff

You'll need this range of tap sizes to clean up all the block threads. Critically torqued threads such as head and main bearing bolts need to be cleaned with a quality tap in as-new condition.

ning at the center cap, torque down one cap at a time and check that the crank spins freely at each stage. When all caps are done, if the crank spins freely, then you can be fairly certain your money spent on line-honing the main bearings is not going to produce any worthwhile benefits.

However, if any noticeable increase in tightness is apparent as the caps are torqued down, then figure some adjustments are going to be required.

Obviously any machining done on the main-bearing bores is going to take metal out and make the holes bigger. To offset this problem, grind the main-bearing caps on the split-line face by a few thousandths inch to close up the housing bore. The line-honing operation will then resize the housings back to the original diameter.

Be aware, though, that after this operation is performed the main-bearing centerline will have moved up into the block. Depending upon how much was taken off the caps and how

After all holes in the block have been rethreaded, a chamfering tool should be used to chamfer off the first thread.

The deck plate introduces the same stresses during honing as the cylinder head does and distorts the block to the shape it will be in operation so that the bores are round in service.

High-Performance Bore Preparation

If not done properly, bore preparation can reduce power and increase oil and fuel consumption. Don't necessarily assume that the machine shop will automatically do the bores properly. Many machine shops simply recondition engines—performance is not a factor. You need to patronize a machine shop that specializes in bore finishes for high-performance engines.

First, the bores must always be machined with a deck plate torqued onto it and the main caps tightened. The boring machine should size the bores 0.0025–0.003in less than the finished size. The finishing operation should preferably be done on an automatic hone such as the Sunnen CK-10.

The type of bore finish used will depend upon the rings. Chrome or stainless-steel rings require a coarser finish than molybdenum rings. Typically a chrome or stainless ring will require the use of a 280 grit stone. So long as cutting pressures are kept to a minimum and the block is flooded with clean, fresh oil, then after four or five final polishing passes up and down the bore, a finish between about 16–25 micro inches should be produced. For a moly ring, a 400 grit stone should be used. This should produce a finish between 12–20 micro inches.

After honing, the bores should be thoroughly cleaned to ensure removal of minute grit particles held in the surface finish. Most of the top piston manufacturers recommend a solution of hot, soapy water and a nylon-type scrub brush. The bores should be scrubbed until there is no more apparent discoloration of the water from any grit being removed. After this, the bores need to be wiped down with white paper towels. When the towels remain white you know the block is clean. After cleaning, either oil or, in dry climates, spray the block with WD-40 to prevent rusting.

much the centerline moved, it may be necessary to use a cam timing sprocket with a bigger pitch circle diameter to take up any chain slack caused by moving the crank center nearer the cam center. Cloyes True Roller timing chain kits made by Cloyes Gear & Products, Inc., are available with oversized sprockets for blocks that have been aligned in the mains.

If realignment of the main bearing is necessary, then this should be the first operation on the block, any quality block rebore job will use the main bearings as a location so bores can be corrected for alignment by the boring machine. Many machine shops are not able to bore blocks in this manner, but for a street motor it's not the end of the world.

If a race engine is being prepped, bore position with reference to the crank centerline should be held to close limits. A bore that is off across the engine has the effect of altering the cam timing in relation to piston position on that cylinder. For the small-block Chevy at least, we tend to think in terms of putting the cylinder bores directly over the crankshaft axis. When it gets down

to it, however, there may be some small justification for actually offsetting the bores—a subject dealt with in the next chapter. For the sort of applications we are dealing with, the bores should be centered over the crank.

After boring to within about 0.002–0.003in, the block must be honed. The least expensive boring and honing job entails just these two operations, but if it is to be done properly for a high-performance engine it is worthwhile paying the additional cost to have your machine shop hone the block with a torque plate in position, especially if you are using a 400 block.

A torque plate for a small-block Chevy is something that any high-performance machine shop should have. Basically its function is to simulate the stresses induced by torquing down the cylinder head; if you're using aluminum cylinder heads, some machine shops will use an aluminum torque plate. The torque plate, together with a gasket, is bolted to the block to induce the same stress as the cylinder head does. Although the block may look to be a sturdy casting, the forces produced by torquing a head down on it are sufficient to distort the top of the block.

What is required is a block that has round bores when the head is on

rather than off, since distortion due to the head bolts occurs right at the top of the block, the most critical area for ring seal. It is often quoted that 80 percent of the power made on the power stroke is in the first 20 percent of the stroke and this seems to be true. If gases escape by the top ring in the first 20 percent of the stroke they will have more negative effect on power than anywhere else in the stroke. This argument emphasizes the importance of honing the block with a torque plate.

Some machine shops run near-boiling water through a block during the honing operation to simulate the thermal stresses involved in a running engine. In practice, the heated blocks don't seem to return better results than unheated ones, however. It could be that the thermal stresses involved by simply running water at a constant temperature through the block don't simulate closely enough what happens in real life.

Under operating conditions the top of the block gets considerably hotter than the bottom, and attempts have been made to compensate for this by honing the block with slightly less clearance in the first 0.50in of the bore than that farther on down. The thought here is that as the block expands, the tapered part of the top of the block will become parallel due to its greater expansion.

If you're working on a budget the next most important operation after boring is decking the block to minimize piston deck height.

After basic preparation and a coat of paint, your block should be all ready for assembly. This will entail putting in the cam bearings, freeze and galley plugs and so on, but before anything goes into the block, make sure that it is 100 percent clean.

High-Performance Engine Block Preparation

2

Block prepping for a maximum-output high-performance engine goes many steps further than the initial preparation discussed in chapter 1. First let's deal with the strength of the cylinder walls.

High gas pressures can put tremendous loads on cylinder walls and any undue flex will result in increased blowby and reduced power. Though rare, flex in cylinder walls can lead to fatigue cracks. If you have the budget, the best plan is to use a Bow-Tie block. Both the metallurgy and increased section thicknesses make this block more suitable for high-output application than a production block.

The Bow-Tie block is a siamese-bore design with no water between the cylinders because they are joined, which has mixed blessings—unless the modifications detailed in this chapter are performed. The main advantage of the Bow-Tie block is that sizing it close to 4.000in bore leaves much thicker than normal cylinder walls. Although it will stand boring to 4.156in, the Bow-Tie block is probably most used at 4.000–4.030in bore.

From the factory, the block is only partially machined and whether you decide to go the Bow-Tie route or to use one of the better production blocks, a considerable amount of work will be required to finish the block.

Casting Inconsistencies

When demanding the best in block machining accuracy, certain dilemmas can present themselves. The problem hinges on the fact that sand castings aren't precise items, and core shifts during production can mean varying sectional thicknesses. Unless you or your machinist is in a position—with the aid of a sonic tester—to hand-pick a block, be it production or Bow-Tie, then there are certain areas where you may have to accept less than the best.

You may well ask where these inaccuracies come from, and why the blocks differ. The answer is simple: the blocks aren't all cast at the same location, from the same patterns, nor are they machined on the same production line. When blocks are cast, the actual process can bring about its own variations.

Pouring near-white-hot molten iron into a mold made mostly of sand and resin is something of a thermal and mechanical shock. Three things happen: particles of sand can break away from the mold and get carried into certain areas; the resins release gases; and the cores themselves, at least those that are installed into the block, tend to want to float in the much heavier iron. Thus the casting process is not totally consistent.

Core shifts and such are bad enough to start with, but production-line inconsistencies compound the problem. Though usually minor, they are relevant. One machining chip in the wrong place can cause a block location error when transferring from one machining operation to another. The result is a less-than-satisfactory piece, even though all the other blocks may have been perfect.

Four-Bolt Main Caps

The most likely operation to perform on many blocks is the installation of steel four-bolt mains caps for the center three caps. Although stock four-bolt blocks are good, it is better to start with a two-bolt and install

A good substitute for a factory four-bolt main block is to start off with a two-bolt block and convert it using steel caps. Here I am prepping a block prior to starting the machine work to install the four-bolt caps.

Four-bolt caps are normally installed so that 0.001in clearance exists between the outer part of the cap and the block when the two inner bolts are tightened down. Because of variations in production tolerances it is necessary to measure the depth of each register with a suitably accurate depth micrometer.

splayed-outer-bolt four-bolt caps such as those produced by Summers Bros. The angled outer bolts produce greater rigidity and a reduced tendency to crack the block.

For most wet-sump applications the stock rear cap is retained for mounting the oil pump. This cap is relatively troublefree unless high output and rpm are envisaged. To beef-up the rear cap and oil pump assembly, mill down the rear cap and install an oil pump-cap reinforcing plate such as sold by Pro Cam.

If you're using a dry-sump system with an external pump, then the rear main cap need have no provision for the oil pump, which makes it easier to produce. This being the case, it is better to install a steel rear main cap for greater reliability.

If steel caps are installed there is usually too much material to be honed out of the now-stepped bore. This means that line-boring must be done. In many instances a good machinist can set up the tooling so that it barely kisses the main-bearing housings in the block while taking the requisite amount from the caps. Realigning the housings in this way leaves the crank-to-cam center distance unaltered and increases the likelihood of a stock gear set giving the correct installed tension—though it is not assured.

Apart from good alignment, the bearing housings need to be accurately sized to produce the correct bearing crush. Middle to lower limit on size is desirable for good retention and heat conductivity. It is also a good idea to check the alignment of the bearing oil holes with those in the block. Any misalignment should be noted so it can be fixed later when

With the cap torqued down on the center two bolts, the outer bolt holes are drilled.

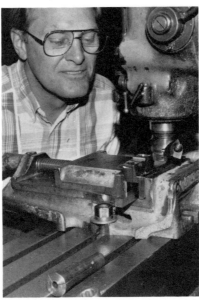

The next operation on the mains caps is to machine the register depth to coincide with that required for the station where it will be installed.

After being measured, each station dimension should be marked on the block. A variation of 0.003-0.005in is common.

Now comes the tapping operation on the block. This is best done starting a tap in the milling machine so that it's in true alignment with the drilled hole.

17

other grinding operations are required.

Deck Heights

The small-block deck height is important because the piston's approach to the cylinder-head quench area is critical if detonation is to be suppressed for maximum power. If crank throws, rod lengths and piston heights are accurately controlled, then the last remaining ingredient for accurate piston-to-quench-area clearance without individually fitting each assembly to its cylinder is to precisely machine the block height.

Both decks must obviously be parallel to the crank centerline, and each deck must be 90deg. to the other and 45deg. to the cam-crank centerline. BHJ makes a special fixture to facilitate machining of the deck and positioning it properly in relation to the crank.

Bore Centering

After the deck is machined, another BHJ fixture can be used to position the bores precisely over the crank centerline and the correct distance from the front face of the block. This ensures that every piston and rod assembly is in the correct position over its relevant crank throw to minimize any side loading.

The BHJ Bore-True fixture also allows you to relocate the cylinder-head dowel holes. Prior to decking the block, tap the cylinder-head dowel holes and insert a threaded cast-iron plug. This gets machined flush with the block deck during the decking operation. With the BHJ Bore-True fixture in place the boring machine locates from the fixture bores rather than the block bores.

In the past there was controversy as to whether or not bores should be directly over the crank centerline. Some engines, such as the flathead Ford V-8, actually used a cylinder-to-crank centerline offset. What this did was increase the proportion of the crankshaft revolution that took part on the power stroke. On some long-stroke engines, offsets as large as 0.250in were used. Of course, offsets such as this are impractical on a small-block Chevy. However, small offsets toward the direction of crank rotation can improve the transmission of cylinder pressure to the crank via a more favorable rod angle.

The indications are that small offsets can provide some benefit. When offsetting a bore, the intention is to relocate the position of the wrist pin in relation to the crank centerline. Again pros and cons exist, but small offsets can be readily accomplished by offsetting the pin in the piston.

Cam Bearing Line-Boring

Inaccurate positioning of the camshaft and lifter bores in relation to the crank centerline is potentially one of the most power-robbing errors. If the lifter bores or cam are displaced in the block from their intended position, a change in cam timing will result. Though the main-bearing bores are line-bored at the factory, the cam bearing bores are not.

Cam bearing bores are factory machined in three operations. First, the two end cam-bearing housings are machined, then the two intermediate ones and finally the center one. After this machining is done, cam bearings are installed and line-bored. When a block is stripped and hot tanked, one of the first things to come out of it is the cam bearings. Replacing them with pre-sized bearings can mean that the original alignment, however good or bad, is lost. Though whatever error that occurs is unlikely to cause a cam to physically bind up, those errors that do exist are just the beginning of a trail of misalignments which can lead to a loss of power.

It is possible to have the cam bearings skewed out of alignment with the crankshaft centerline when viewed from above the block. This axis error can cause cam timing to

Next operation is to ream the caps out for clearance on the bolts.

Here's the block with the Summers Brothers' caps installed ready for aligned boring.

vary on each cylinder. Moving the camshaft centerline from one side of the crank centerline to the other will produce timing changes. If we were dealing with an inline engine such errors could be adjusted out, but with a V-8, displacing the cam to one side causes one bank to advance and the other to retard.

When the cam is timed in on number one cylinder, any error that may have been split between each bank is now applied solely to the other bank. This can drop output more than if the error was split evenly between each bank. Still, errors at this point are normally minor and can be partially compensated for by checking the cam

timing on number one and number two cylinders and splitting the error.

Positional displacement of the lifter bores is more difficult to deal with and cannot be compensated for by cam timing because the lifter bores may not exhibit the same errors down the length of the engine. Some lifter bores may produce advanced timing and others retarded, depending on the *angle* they make with the cam. Others may be displaced such that the cam and lifter centerlines don't intersect and, as far as the cam is concerned, the lifter diameter is effectively reduced. To avoid edge-riding of the lifter on the cam lobe, a less-aggressive cam profile must be used, which is not desirable for a maximum-output engine. The only sure way of correcting the problem is to blueprint the crank-cam-lifter system.

To correct cam-lifter centerline errors, begin by line-boring the cam

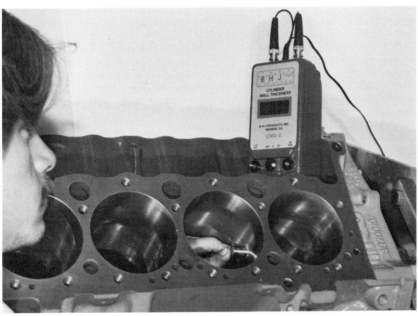

Core shifts make for variation in cylinder wall thickness. Sonic testing on this block showed that the cylinder walls on one bank of cylinders were a little on the thin side. A block like this may be OK for a regular street motor, but certainly not for a high-performance engine.

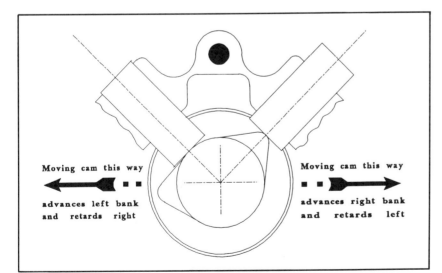

This end view of the cam and lifters indicates how an inaccuracy in the cam location can affect the timing; however, this is just a simple case. The alignment of the cam may not be perfect, and it could be off at an angle relative to the crank. This causes an advanced error at one end of the block and a retarded error at the other.

Though line-honing will work for truing up stock caps, the installation of steel caps, because of the greater amount of material to be removed, requires line-boring.

19

bearing housings and installing the Sealed Power cam bearings with oversize OD. During this operation, some engine builders increase the size of the oil groove in the cam bearing housing to improve oil flow to the mains.

Lifter Bore Centering

With the crank and cam as reference points, lifter bores can be positionally corrected in relation to the cam centerline. There was a time when this was a lengthy operation for a milling machine, but BHJ manufactures a fixture that simplifies precision fitting of the lifter bores in relation to the cam on a small-block Chevy. This fixture uses two special end plates, which are positioned off mandrells located in the cam and main bearings. The cam location mandrell has pilot holes corresponding to the lifter centerlines. This, in conjunction with an accurately located plate above the lifter valley, which also has pilot holes, allows a cutter to reposition the lifter bores with far more precision than the original factory bores.

It is not uncommon for the lifter bores to be displaced as much as 0.020in, and accumulated errors can alter valve events by as much as 7deg. If you consider that lifter bores can have more or less at random errors, then it's not hard to see that any timing precision on the cam is lost as the motion is transmitted to the valve. Blueprinting the lifter bores remedies this, but boring them means they are now oversize.

The standard diameter for resizing a small-block Chevy lifter bore is usually 0.875in, the stock size of a Ford lifter. Some racing rules call for stock 0.842in diameter lifter size. To compensate, BHJ produces sleeves that can be used to resize back to the stock Chevrolet dimensions.

The bottom line is that bigger lifter diameters mean higher maximum lifter velocities. If there is no reason to limit the size of the lifter, then it's a good idea to leave it at 0.875in diameter. This typically allows lifter velocity to be increased from 0.007525in per degree to 0.007612in per degree.

Correcting the position of the lifter bores is one of the most exacting jobs to be done on a small-block Chevy and, aside from obvious blueprinting exercises to minimize friction, on average it produces the best returns for the effort involved.

Assuming the pistons, rods and crank are dimensionally accurate, then to get a minimum piston-to-deck clearance, it's necessary to have the block decked parallel with the main bearings. This BHJ fixture ensures that the block will be both parallel with the mains and that the decks are square to each other.

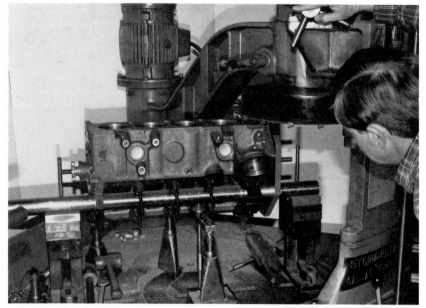

Here's the BHJ Block Tru set up ready to start machining. For the most part it's convenient to deck small-block Chevys to a 9.0in crank center to deck height.

This BHJ Bore Tru fixture is designed to locate the bores directly over the crank centerline.

The last block modification concerns the main and rod journal oiling system. At high rpm there is evidence that oil flow to the mains from the main galley in the middle of the V is restrictive. The first step towards rectifying this is to drill the oil feed from the main bearing to the galley about 0.030-0.050in larger.

At this point the groove behind the cam bearing becomes the limiting factor. This can be machined out to good effect but at the present time I feel that the majority of the restriction here is caused by the poor flow characteristics of the junction of the oil hole drilling with the cam groove. Much of the effect of costly cam groove machining can be obtained by flowing the transition points.

Supercharged and Turbocharged Motors

If you intend to supercharge your engine in any way, shape, or form, a few precautions and modifications here can help the reliability of the engine substantially. Firstly, the pistons in a supercharged engine are subject to far higher thermal loadings. This means they either need to be thicker, to be able to cope with their reduced material strength they will have at the higher temperatures involved, or they need to be cooled to keep them operating in a higher strength temperature range. The easiest way to do this is to put oil cooling jets into the main caps of the block, the nearby photos show how this is done. Essentially it entails grooving the main bearing saddle in the block so that oil can flow from the oiling hole in the main bearing saddle to newly drilled oil squirt holes. These need to be aimed at a slight angle

To true up lifter bores this BHJ Lifter Tru fixture is used. The first move is to install this bar with accurately positioned holes into the cam bearings.

Next, the end plates are installed. This puts the cam bar in the correct angular orientation.

Once the top plate of the Bore Tru fixture is located it is possible not only to accurately position the bores, but also to reposition the cylinder head locating dowl holes in relation to the bores.

towards the base of the piston. The best way to visualize this is that if the piston is halfway down the bore, the jet of oil will be aimed at the middle of the crown on the inside of the bank. As the piston travels up and down the bore, so this jet of oil will sweep across the piston, spread out over the underside of the crown of the piston, and will be thrown off from the other side. The size of the jet is important. Obviously, if too large a jet is used, the oil pump simply won't keep up with the demand, on the other hand, too small and the cooling effect won't be achieved. Basically, the big block Chevrolet oil pump can supply oil for cooling through 0.0028in jets for each cylinder as well as the oil demand for the rest of the engine. Don't attempt to use these oil jets with standard output small block pumps, they're just not up to it. Also, be sure to note that flow restriction at the cam bearings is now more important to cure, because of the additional demand made by the cooling jets.

This allows location of the top plate. The accurate holes in this plate and the cam bar now provide an accurate reference for the cutter to true up the cam follower holes.

Using a suitable machine bar located in both the cam bearings and main bearing housings, the cranked cam center-to- center length can be determined. **High Performance Communications**

From this dimension a suitable gear set can be selected using this BHJ gear set center-to-center measuring fixture. **High Performance Communications**

The oil for cooling pistons for supercharged applications can be drawn from the main bearing supply hole by grooving and drilling as indicated here.

An oil hole here cuts wear on the cam sprocket and block thrust face. Also a very small bleed hole—just as small as you can make it—in the galley plug above will vent trapped air, thus preventing a short no-oiling period at startup at the front main and rod bearings.

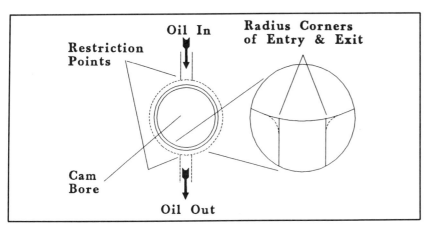

Here's a simple view of the passage of the oil into and out of the cam bearing groove. The greatest restriction points are indicated by the arrows. This can be rectified by grinding as detailed here.

The amount of oil to the underside of the pistons for cooling purposes is controlled by a 0.028in hole drilled in the pressed-in brass jet. When this modification is done, be sure to maximize oil flow around the cam.

Race and Big-Inch Motor Preparation

3

You've heard the saying, There's no substitute for cubic inches. In essence, it's true. Yet it is not without compromise as far as the small-block Chevrolet is concerned. No production engine can be stretched to unlimited cubic inches; there comes a time when boring for more cubic inches becomes dubious. As always, one must consider the intended application.

If you plan to build a high-output street motor, then cubic inches is as good a way to go as any to get street power. A street motor should also have a high degree of reliability, again depending on its intended purpose. If an engine is expected to go 50,000 miles, the specification needs to reflect this. On the other hand, if the motor needs to hang together for 1,000 miles on a Baja dash, then wearing out at 1,100 miles is no big deal as long as your budget can afford a new engine every race.

The limiting factor is that a small-block Chevrolet has a limited deck height, and increasing the stroke of the engine almost inevitably means limiting the rod-to-stroke length ratio. When the length of the connecting rod is shortened in relation to the length of the stroke, piston and cylinder wall side loading is increased. As a result, pistons and cylinder walls can suffer accelerated wear.

Though small-block Chevy engines as big as 440ci have been built, the questions arise as to whether they are worth the time and effort, whether they are reliable and whether the torque and horsepower produced warrant the required effort. Remember, as stroke increases, so does bore friction and if the engine is limited by cylinder-head airflow, as most small-block Chevys are, then it could mean that friction increases at a faster rate than air consumption at more than 5000rpm. The consequence is that maximum horsepower does not significantly change.

On the other hand, bigger cubic inches almost always deliver better low-end torque but if rod angularity is excessive, which it will be, then even the bottom-end torque may not be all that is expected.

At what point do the returns start to outweigh the disadvantages? It is difficult to put a concrete figure on this, but around 430ci seems to be the cut-off size for a small-block. At any rate, prepping a 400 or siamesed-bore block can get expensive, so let's begin a little farther down the scale.

350 Block

If you're building a motor from scratch there is no point in utilizing anything less than a 350 as a starting point because short motors are virtually a dime a dozen. Considering cooling, the 350 block is a more reliable piece to work with than a 400 block, and is more plentiful. Motors of 400ci tend to wear out blocks—especially the back cylinders—causing them to be unboreable, even to 0.030in over. Bore sizes much bigger than that on a 400 block can spell trouble with flexible bores. Since the blocks are less reliable than crankshafts, this often means there is a surplus of 400 cranks.

Since the 400 crankshaft has a bigger main-bearing diameter than a 350, you can grind down this bearing to a 350 size to utilize the crank in a 350 block. Increasing the stroke from the 3.480in used on a 350 engine to 3.750in used on a 400 stretches a 4.030in (including a 0.030in overbore) bore 350ci block to 383ci for a 9 percent increase in capacity. The conversion is inexpensive to do because cast-iron cranks are relatively low in cost, and if performed correctly it is an effective way to achieve a good, torquey street motor. From mid 1985 onward, building 383ci motors has become very popular and they represent a sizable proportion of hopped-up Chevrolets.

In spite of this, most 383s built suffer from some elementary mistakes that invalidate the potential gains of the extra 33ci. The easy way to build a 383 is to use the 5.56in long 400 connecting rod together with a regular off-the-shelf flat-top piston and an open-chamber smog or low-compression cylinder head. This combination produces little or no more horsepower than a typical 350 built with a long rod and closed-chamber head. It does use more air and fuel, though, meaning it is an inefficient package. If you are going to build a 383, do yourself a favor and do it properly. It will cost a little more, but it will be money well spent.

First, do not use anything less than a 5.70in rod—that's a 350 rod, for your 383. Because this type of conversion has gained so much popularity there are, as of 1991, pistons made specifically for building 383s from a 350 block using this rod length. These pistons have a shorter pin-to-deck height so the piston does not come out of the top of the block. Also, there are pistons available with a variety of crown shapes to give the desired compression with almost any configuration of cylinder head you are likely to use.

The closed-chamber high-performance head—the 186 casting is typical—is a good choice. If you are going to build a 383 using a 350 block with a ground-down 400 crank, both the rods and the block will need a bit of clearancing. The rods will just hit the sides of the block, and it requires about 0.125in ground out at the appropriate places.

Depending upon the tolerances, the rods themselves can either just clear or hit the flanks of one or two of the cam lobes. There are two ways around this. Either use a cam with a 0.050in small-base circle diameter, or grind the connecting rod in the area of the bolt head, as described later. Many people are reluctant to grind the rod, but practice has shown that no significant weakening results if the rod is ground properly. On the other hand, grinding a camshaft with a 0.050in smaller base circle may produce a cam with a softer surface than would otherwise be the case, and the cam may wear itself out prematurely. I prefer to go with the rod-grinding technique.

Bow-Tie and 400 Blocks

The Bow-Tie and 400 blocks are siamesed-bore blocks without a water jacket between the cylinders. The castings of the outer wall of the cylinders actually join along the centerline of the block because they are so close. This produces a more rigid block, but also introduces some inherent cooling problems. These are correctible to a large degree, but not without some hassle.

As it comes from the factory, the Bow-Tie block has main-bearing saddles sized for the 350; the 400 block has larger main bearings. For most hot rodders the 400 block represents a less expensive alternative to the Bow-Tie block for building a large-displacement engine. Although it may be difficult to find a 400 block that can be bored 0.030in over, this route still proves less expensive than a Bow-Tie.

Let's begin at the smallest displacement likely to be required using a 400 block. Many racing rules call for a 355 or 358ci capacity limit. If there is no restriction on the use of a stock bore and stroke, it is possible to build a short-stroke, high-rpm 350. Because most of the mechanical losses in an engine are the result of piston and ring friction, cutting the stroke length means less power loss in this area. Also, because of the larger piston area big-bore short-stroke engines tend to make more horsepower per cubic inch than do those with a smaller piston area and longer stroke. By using a stroke of around 3.250in (almost 0.250in shorter than a 350) and a 4.156in bore, we are able to build a 350ci short-stroke engine. The shorter stroke allows the use of a much longer connecting rod that in turn permits cylinder wall loadings to be decreased to further enhance the engine's potential power.

There is another alternative, however, especially if there are budget constraints. The combined use of a 350 crankshaft and a 400 block will produce around 377ci and, in essence, is like having a big-bore 350. This engine makes a good combination because the extra capacity has been achieved by virtue of a bigger bore rather than a longer stroke. As opposed to building a 383, the problem is finding a good block as a basis. With a suitably prepped stock forged crank, a 377ci small-block is much more of a high-rpm unit than a 383. All other things equal, it is capable of delivering more horsepower and is a

Here's another area that can be usefully blended out on a long-stroke big bore motor.

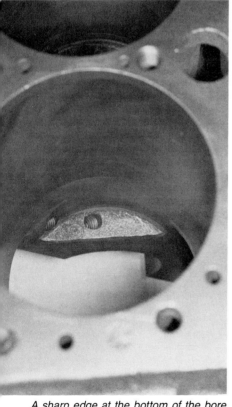

A sharp edge at the bottom of the bore between the bore and top of the main-bearing housing can be a source of cracks. Polishing out this corner can reduce this.

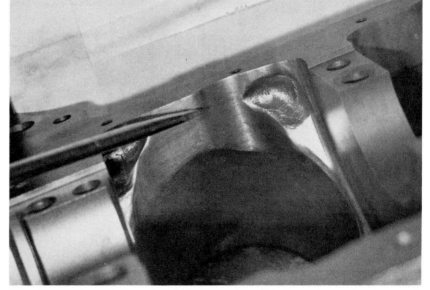

When long strokes or bulky rods are used, it is necessary to grind some rod clearance in the block. Note also how the area adjacent to the main cap has been polished out, again in an effort to reduce cracks.

If the block is to be O-ringed, this steam hole needs to be tapped appropriately and have a soft iron plug installed.

noticeably more free-running configuration.

To use the 350 crank in a 400 block some special bearing spacers are required. These inserts install into the larger 400 main-bearing saddles to allow for the use of regular 350 main bearings; companies such as RHS carry these bearing spacers. Another popular modification was to use special thick-walled bearings such as produced by TRW but these are far more expensive than regular bearings, so it becomes a costly procedure to replace bearings. Resizing the bearing inserts to 350 proves much cheaper when it comes to bearing replacement.

If a good 400 motor is used as a basis for a rebuild, then you're likely to increase cubic inches relatively easily. Boring the block 0.030in over—it will almost certainly need a rebore unless you begin with a new block—will result in 406ci. Again, it's not a good idea to use the 5.56in rod unless the application is for a mild street motor. Another technique for stretching a few more cubic inches out of the 400 block using relatively stock parts is to pair a 400 crank with 5.70in 327 small-journal rods. The 400 crank big end can be offset-ground to increase its stroke slightly. Together with a 0.030in overbore, this can produce 415ci.

I've had no direct experience to speak for the reliability of the 400 crankshaft with the smaller crankpin. It is reasonable to assume, though, that if a good crankshaft damper is used and the fillet radii in the corners of the crank are correctly prepared, and maybe some suitable heat treatment has been done on the crank, then the restroked 400 crank would be at least as reliable as a stock unprepared crank at standard size up to 6500rpm. To get more than about 415ci from the small-block it is necessary to look at longer strokes from the crankshaft. Bore sizes much over 4.156in are not practical. With a 4.156in bore and a 4.000in stroke the small-block Chevy can displace 434ci.

Big-Inch Block Preparation

All this sounds good, but making a reliable big-inch engine is more complex and expensive. The procedure I will describe is based on what's required to prep a 400ci block; much of this will also apply to a Bow-Tie block, except for details relating to the main bearings, which may not need replacement.

First, if you intend using a 400 block, then the casting number to look for is 509 cast into the side. Failing this, any of the others will do.

If you are working on a relatively small budget and you feel the en-

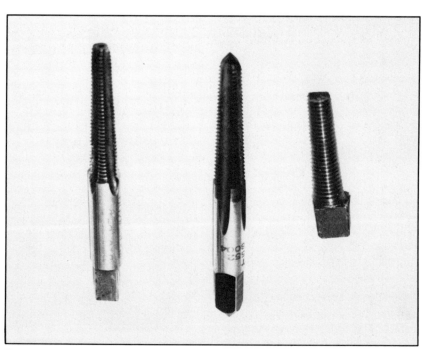

This steam hole also needs to be tapped if you are going to O-ring the block and have the soft iron plug installed.

This is what will be needed to tap and plug the 400 block steamholes.

From the groove toward the edge of the fire ring in this gasket you can see where the wire O-ring embeds.

gine's output will warrant it, the factory four-bolt setup is worth looking into. At high outputs, four-bolt blocks tend to crack immediately below the outer bolts and this is why the splay-bolt-pattern cap proves superior. However, if your budget stretches to include installing mains caps on a two-bolt block, this is the best way to go. Most 509 blocks will be of this two-bolt configuration.

If your intention is to line-hone a 400 block, note that the smallest diameter of the rear main seal is

If you can find one, the best 400 block to have carries the 509 casting number. 509 blocks were sculpted to remove any unnecessary metal. This is especially noticeable around the fuel pump mounting position. Though I have never made a comparison, it's said to be about 12lb lighter than most other 400 blocks.

Cutting this groove in the block, without the steam holes being plugged, means that the groove crosses a steam hole, as seen here.

After the water holes in the deck face have been plugged, the deck will look like this.

If the engine develops a lot of horse-power such as in a blown or nitro-injected motor, this can happen to a production block, though heavier Bow-Tie blocks are significantly stronger.

The stock rear main cap, especially on a 400, can be prone to cracking. Steel support plates can be obtained from Moroso or Pro Cam to reinforce this area.

Installed, the steel support plates look like this.

smaller than the OD of the bearing housings. Many machine shops will machine this seal OD larger so that the honing stones can pass through the mains. This means that the rear oil seal on the crankshaft will be incorrectly located. To overcome this problem, machine up a ring in any machinable material such as brass, aluminum, mild steel and so on, cut it in two and insert it into the groove in the rear main seal so it replaces the metal that was removed for the honing operation.

On engines producing 600–700hp, especially those for circle-track applications where sustained rpm is needed, the blocks crack at the bottoms of the bores in and around the junction with the main webs. Some of this cracking is caused through flexure at stress-raising corners; this can be delayed if those corners are rounded off. A little grinding in and around the area of the main-bearing webs in the block can help reduce the incidence of cracking.

Modifications also need to be made at the deck of the block, especially if the block is to have O-rings installed. These modifications are designed to stabilize the top of the bores and to change the way the water passes through the block. If the block is to be O-ringed with the O-ring falling toward the edge or even just outside the fire ring of the gasket, the O-ring groove will cut into one of the steam holes between the cylinders. To fix this the steam holes need to be tapped and a taper iron plug installed. The length of the taper plug should be limited to a little less than the depth of the deck because another drilling operation must be performed later in which excess plug protrusion from the deck would be undesirable.

The large water holes in the block face need to be tapped with a plug tap, and a suitable cast plug installed at this point. One function of this plug is to cut down or seal off completely the size of the water hole. If the water hole is to be redrilled, it can be to the size of the gasket, which is usually restricted to a relatively small hole anyway. The plug's other function is to pre-stress the block outward from the hole so that combustion pressures do not cause the top of the cylinder bore to move quite as much. The presence of this plug seems to reduce the amount of distortion resulting when a torque plate is installed and you can expect the cylinder bores to retain

their roundness better under operating conditions.

Having blocked the steam holes, there seems to be mixed opinions as to the usefulness of redrilling these. I've redrilled steam holes at a smaller size to clear any O-rings that may have been installed. In addition, on 400 and Bow-Tie blocks I've drilled an interconnecting hole between the cylinders from the lower side of the water jacket to the upper side immediately below the deck face. This allows air or steam that may become trapped in the pocket on the underside to escape to the upper side and be carried away by the coolant. If a steam pocket forms on the underside it prevents cooling at that point, causing higher temperatures that generate more steam and even overheating.

Although this cycle of events can lead to problems, it doesn't usually destroy the engine unless it causes the water to be blown out of the radiator. If it does, then the motor won't be long for this world. If the steam holes are plugged it would seem necessary to drill the horizontal hole under the deck. If we are dealing with a 400 street motor with relatively high output, then it is only necessary to drill a hole between the center two cylinders on each bank. The reason is that at this point two exhaust valves are adjacent, resulting in far hotter running temperatures than other areas, hence the importance of ensuring that no steam pocket forms.

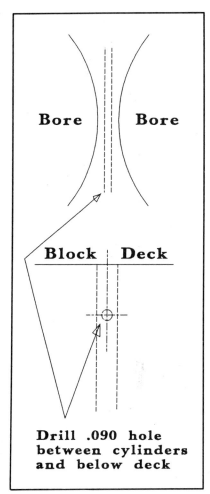

Though it's a tricky job, drilling a 0.090in hole between the bores on a siamese block, as detailed here, helps keep gaskets intact and the top of the cylinder a little cooler.

Chevrolet Pistons Selection and Preparation

4

Pistons are probably the most abused components of a high-performance engine. They are subjected not only to substantial reciprocating loads, but also to high gas pressures and temperatures. All these conspire to make the life of the piston designer difficult. If we add to the cauldron the necessity of making pistons light, quiet running, durable, scuff and seizure resistant and so on, we can see there is more to piston design than just a simple slug of aluminum with rings and a wrist pin.

Although factors may change in the not-so-distant future, the manufacture of pistons currently falls into two broad categories: cast and forged. In addition, each of these categories has a subcategory of its own. For cast pistons a conventional casting material can be used, or they can fall into the hypereutectic category, explained later.

Forged pistons are generally termed low silicon or high silicon. In all instances, the amount of silicon alloyed into the base aluminum is the differentiating factor. We'll deal with the reasons for its use in a moment.

Cast Pistons

For the most part, stock pistons are cast with a steel insert by each wrist pin boss to control expansion. For a regular street motor, power output is not necessarily of prime importance but what is vital is long life and quiet running. For a production street engine, the piston manufacturers attempt to produce pistons that have a low expansion rate so when the engine is first started in the morning there is minimal to zero piston rattle prior to reaching operating temperature. Additionally, the pistons are required to run with a close clearance so the engine is quiet. Though this type of piston will take a reasonable power output, it was never intended to withstand the high output of a race engine or even a relatively high output street motor.

For a typical small-block Chevrolet street motor 0.85hp per cubic inch—that's 300hp from a 350—is about the upper limit for a cast, steel strut piston. That doesn't mean this figure cannot be exceeded, however. If an engine is built with significantly more power than this it had better be with a piston that is a known quantity in terms of its ability to take such power, and that the time the engine spends at such a level is limited.

For a regular rebuild and modest power outputs at a reasonable price a cast piston may not be all that bad, but there are alternatives. In the mid 1980s a new type of cast piston known as a hypereutectic piston came to the fore. The term hypereutectic indicates that the major alloying substance employed is in excess of the amount that will dissolve in the parent material. In this particular case the alloying substance is silicon.

Most grades of aluminum will only dissolve about 12.5 percent silicon, and most cast pistons have about 9.5 percent silicon; a hypereutectic piston is one which employs greater than 12.5 percent silicon. For most hyper-

eutectic pistons currently available, the quantity of silicon can range from 16 to a little over 20 percent.

When discussing silicon we are, to put it simply, talking about sand. The positive aspect of adding silicon to aluminum is that it becomes harder and more scuff resistant. On the debit side we find that the material becomes more brittle and consequently, more prone to cracking. However, this isn't a problem of any magnitude with the amounts of silicon normally used in pistons. Indeed, most pistons still retain more ductility than necessary.

Another advantage of high-silicon content is reduced heat conductivity. This means more of the heat in the combustion chamber remains there instead of being conducted through the piston crown. There are pluses and minuses for this as well. Heat that remains in the chamber means more power but if the surface of the piston runs hotter, then detonation fast becomes a factor to contend with. Another major asset of high-silicon-content pistons is that the silicon reduces the expansion coefficient of the pistons.

Hypereutectic pistons are made solely of aluminum alloy rather than having a steel strut incorporated. Due to the hypereutectic pistons' advanced technology, they are rapidly becoming popular for high-performance street use. Such pistons can be fitted with close clearances for quiet running, yet they can handle relatively high power outputs and still deliver extended ring groove and skirt life.

These pistons are typically good to 450hp and in some instances as much as 500hp in a small-block Chevrolet. Power limitations this high should cover most street applications. They are a little more expensive than the regular cast piston, but are generally less expensive than their forged counterparts.

Forged Pistons

The difference between a forged and cast piston is that a cast piston is poured from molten metal into a

During the mid-1980s the hypereutectic piston, so-called because of its high silicon content, became popular. Its low expansion means close clearances, and good hot strength allows it to be used even in highly modified street motors.

30

mold. A forged piston begins life as a billet which is put into a die and then stamped into the basic form of a piston by a punch. Basically the most popular pistons—that is, TRW and Sealed Power—are forged at temperatures around 900deg. Fahrenheit. These pistons are made of an alloy known as MS-75, which contains 11 percent silicon specially developed for TRW for piston manufacture. This high-silicon alloy brings some of the same advantages high-silicon content brings to cast pistons. With 11 percent silicon it isn't hypereutectic, but the silicon does improve scuff resistance and ring groove life by increasing hardness.

Most other piston manufacturers, including most custom piston manufacturers, tend to use alloys with lower silicon content, often as low as 0.1 percent silicon. But this need not be all that unfavorable. A lower silicon content can increase the toughness of the piston and, assuming the correct type of alloy is chosen, heat-treatment processes applied after the piston is rough-machined bring the hardness of the alloy up to adequate levels. Granted, the effect of the heat treatment will be lost on the crown due to combustion temperatures. However, most of the hardness is required on the skirt and in the ring lands and this is where it is retained.

What are the advantages of a forged piston? The prime factor is piston strength. A forged piston represents the ultimate way to go for a high-output motor, though hypereutectic pistons are not that far behind.

For street use you need to carefully consider whether or not a forged piston is necessary. If nitrous oxide injection is to be used or if the engine is to be twisted at high rpm and near-detonation levels for any length of time, then a forged piston is justified. If compression ratios are being run near the ragged edge, then a forged piston usually has a better margin of safety against detonation than its cast counterpart.

Some forged piston manufacturers claim that the grain flow of a forged piston, which is oriented during the forging procedure, is an advantage. This may appear so at first glance but it seems conclusive, especially at the temperatures a normal race engine operates, that oriented grain flow is worth nothing.

Piston Clearance

One of the main factors to understand about pistons is the clearance required to compensate for expansion. Some of the cast pistons can run with small clearances, while some of the forged pistons require large clearances. Large clearances would seem a disadvantage only while the engine is cold because as the piston is warmed up it will expand to operating temperature and fill the bore in just the same way the tighter clearance piston does with a lesser expansion coefficient. Yet even though we may make allowances for piston expansion, for the most part, a piston with minimum expansion is considered an asset.

The problem with allowing the piston to expand to fit the bore is that one must assume it will only be operating at one particular temperature—which is not the case. To understand why, let's consider a piston that has no expansion whatsoever. Under these circumstances it can be made to fit the bore precisely, and such a piston will hold the rings tight against the cylinder bore under all conditions, whether hot or cold. This is important for good seal. Consider, also, a piston that expands by a fairly large margin. When the piston is rocking around in the cylinder, the rings are not square against the bore and they will leak. Measurements of blowby show that pistons with large clearances have much more blowby than pistons with tight clearances. Here we are talking about the actual clearance that exists when all the effects of temperature are taken into account. A good rule of thumb for a small-block Chevy seems to be that for every 0.001in of extra clearance,

This TRW piston is a prime example of a popular high-silicon, forged piston.

Low-silicon alloy pistons often used for racing, such as the popular Ross piston seen here, normally need greater cold clearances because the alloy expands to a greater degree. When thermal barrier coatings are applied, the running temperature is reduced, so cold clearances can be cut slightly.

about 1½cfm (cubic feet per minute) of additional blowby is produced from about 0.002in piston clearance and upward.

How does this affect the engine when a high-expansion piston is used? If such an engine is to be used for the street, then obviously the temperature the piston will run at during cruise will be significantly different to that at full throttle. Therefore, at cruise rpm a forged piston with a high coefficient of expansion will experience much more blowby because the piston hasn't grown to fit the bore. If the piston is sized so it is optimal at cruise, then it is likely to seize under full-throttle conditions.

Therefore, a high coefficient of expansion is not something to be side-stepped by simply concluding that the piston grows when the engine gets hot—the amount of heat involved varies, depending upon the engine's use. On the other hand, an engine with a low-expansion-coefficient pistons can fit the cylinder more closely under all conditions. This point must be considered when selecting pistons for a street motor.

If it is a race engine, the situation is different. Part-throttle conditions are not crucial. The only factors worth considering are how reliable the pistons are and how much horsepower they will allow the engine to generate. Under these conditions a high-expansion forged piston is no real problem—except that it's noisy when the engine is first started up. The only concerns are that the pistons fit the bore correctly and control the rings as required. They must also survive under wide-open-throttle conditions when the engine temperature is at race level.

Forged pistons that have a high-silicon content, such as the TRW or Sealed Power, can generally run tighter clearances than those with lower silicon content. Yet even these forged pistons require clearances in the 0.004–0.006in range rather than the 0.0015–0.0035in range required of various types of cast pistons. Pistons with a low-silicon content generally require clearances up around 0.007–0.008in, which of course decreases as the engine warms up.

The amount of cylinder-wall-to-piston-skirt clearance used isn't solely dictated by the piston materials. The piston design also plays an important part in determining exactly what cold clearance a piston can use and what that clearance ends up as when hot. An example of how skirt design can have a measurable influence on the minimum piston-to-cylinder-wall clearance is a split-skirt piston. This type of piston is out of fashion, but it does serve to illustrate the point.

A split-skirt piston has a vertical slot in the skirt; as the piston expands, the slot allows the skirt to be flexible enough to bend in to compensate for the expansion. The result is that these pistons can be run with a close cylinder-to-piston clearance. This of course makes for quiet running, though increased piston friction is the penalty. Split-skirt pistons are no longer in vogue because they simply aren't strong enough for today's engines, which develop much higher pressures than did early post-war engines. However, the split-skirt is not the only way to reduce the amount of cylinder-wall cold clearance. One of the major ways to keep the cylinder-to-skirt clearance to a minimum is to try and isolate the amount of heat dissipated through the ring belt into the skirt.

A piston's oil ring drain-back features are important in controlling skirt heat. There are two chief differences in oil ring groove design: pistons with oil return holes from the oil control ring back into the crankcase drilled through the base of the oil ring groove, and pistons with a slot. The slot puts an air gap between the crown of the piston and the skirt, so the amount of heat conducted to the thrust face of the skirt is reduced, and this is where most of the piston slap is controlled from. Therefore, a piston with a slotted oil control ring groove can run with tighter cold clearances because the skirt expands less due to reduced heat conduction.

If this is so, why don't all pistons have a slot put into the oil control ring groove to isolate the heat? The simple answer is that it weakens the piston. The piston crown is now supported only at the pin bosses instead of

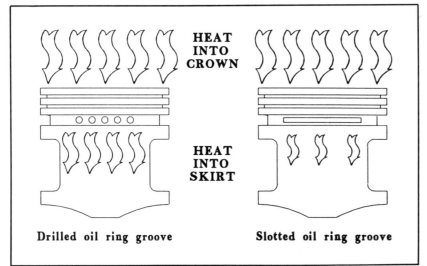

Though the drilled style of piston on the left is stronger, it allows more heat to be transmitted into the skirt than the slotted style on the right. If piston expansion looks like it could be a problem, then for a given material the slotted ring groove allows closer clearances.

Not all race pistons are low-silicon alloy pistons. Forgings from Wiseco, Sealed Power and TRW can all be had in high-silicon alloys, typically of 11 percent silicon.

around the entire periphery. Also, the skirt becomes more flexible because a large portion of the skirt now is unsupported both top and bottom instead of only at the bottom.

For most street applications, though, the slotted oil groove in a sturdy forging is an excellent compromise. If correctly made, there is no reason why a slotted oil groove cannot be used for pistons in engines up to about 450hp. I have used Cosworth pistons at power levels of 1.2hp per cubic inch of displacement. So if you want a good, stout forged piston, then this is one aspect to consider.

If a high-silicon piston is used, it may not be necessary to go to such extremes to isolate heat from the skirt. For instance, Wiseco pistons use a 12 percent silicon alloy and, like the Sealed Power and TRW forged pistons, can be run with clearances down to 0.004in. Indeed, if a careful break-in procedure is used, piston clearances even smaller than this can be used. At clearances down to much less than 0.004in you had better know exactly what you're doing, however.

The form on the piston skirt is also important with regard to the amount of cold clearance used. Piston manufacturers tend not to talk about the barrel and cam forms on their pistons simply because they often regard them as top-secret information. This seems to be taking things too far because if a piston manufacturer wants to know what someone else is doing, it's easy to simply buy a competitor's piston and measure it.

Piston Selection

Even when building a relatively stock motor, you must consider exactly what you need in terms of pistons. Certain prime factors need to be known, such as the required compression ratio. To determine what is necessary in the way of a piston dome, it is essential to know the cylinder-head combustion chamber volume along with the gasket and piston clearance volume in the block. Ideally the block should be decked to give the piston as close to a zero piston-to-head quench area clearance as possible.

For a street motor, the goal is to bring the pistons to within about 0.036-0.040in of the cylinder-head quench area. If it's a race engine that's being built and the parts are of known values in rod stretch, crank flex and so on, then figures a little closer than this can be utilized; clearances as close as 0.030in may be sufficient for certain parts combinations, but if rpm over 7500 are used, static clearances of 0.045in are more appropriate. If aluminum rods are used, the clearances must be increased to about 0.060in because these stretch and expand so much more than steel rods, therefore diminishing the clearance as rpm and temperature increase.

Let's assume the required compression ratio is known and an off-the-shelf piston is available. What other factors should be considered? The first question you should ask is, what rings are best suited to the application? The next points to consider include type of pin fit and overall piston weight.

Ring Width

Most Chevrolet pistons are available with 0.0781in (5/64in) or 0.0625in (1/16in) wide compression rings, whereas many of the specialty pistons required for all-out performance are available with ring widths down to as narrow as 0.043in. For the most part, the narrower rings are preferred for a high-rpm engine; they will seat in quicker and are less likely to flutter.

Ring flutter occurs when the momentum of the ring at the top of the stroke causes it to unseat from the bottom of the groove, allowing the gas pressure from combustion, which is supposed to seat the ring against the bore, to escape around the back of the ring. Since the ring is not seated properly and because the pressure gets to the sealing face first, gases can leak across the sealing face of the ring.

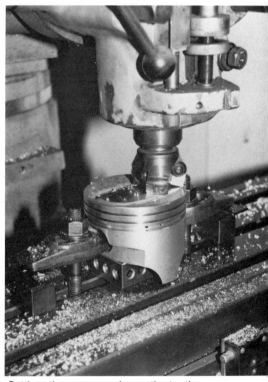

Getting the compression ratio to the required level is important. For the most part it is better to mill the heads to get the minimum chamber volume possible consistent with retaining mechanical integrity of the head and then machine the piston dome so the desired final combustion-chamber volume is achieved.

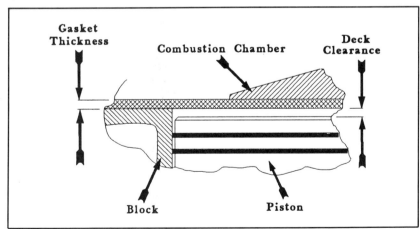

Minimizing the piston-to-quench distance helps power output as well as reducing the tendency to detonate. Most blocks need machining about 0.025in to produce zero deck clearance, but check before assuming 0.025in off will be right. Production tolerances vary on blocks, pistons, rods and cranks, thus making it impossible to make an absolute assumption without previously measuring.

Fortunately, ring flutter is not a prevalent problem for small-block Chevys with correctly designed and made pistons. For this reason, super-narrow rings are not essential.

If we are looking at an engine for high-performance street use, the wider the ring is the longer the ring will last because it has a larger surface area over which to distribute wear. For engines running up to about 6000rpm, the 0.0781in ring is probably the best bet. For engines going to higher rpm than this, say to 8000rpm, the 0.0625in ring gets the job done. For engines turning rpm above this the narrower ring may be better, not necessarily because there is extra power to be gained from the ring itself, but because the sheer geometry of the situation may produce an advantage.

If you are building an engine to turn 8000rpm or more, it is likely to be a serious effort to maximize horsepower—in which case the geometry of the bottom end becomes important. At higher rpm levels, rod length becomes more important. A small-block Chevy has limited space to accommodate the kind of rod lengths normally required. Using a narrower ring means the rings can be closer, allowing the pin to be higher in the piston. This may be a small point but gaining 0.050in rod length and the associated extra skirt length on what could be a short piston, just might add that edge.

Wrist Pins

For most performance rebuilds with mostly stock parts, you need to determine whether or not the wrist pin can accommodate a floating or press-fit setup in the rod. The stock rod employs a press fit, which has some advantages. All things equal, a press-fitted pin is stiffer than a floating pin because the rod helps stiffen the pin's center section. On the other hand, a press-fitted pin is not easy to change and any attempt at disassembly can result in piston damage. Also, wear on the pin-piston assembly occurs only at the piston-pin interface;

a floating pin distributes wear over three bearing surfaces instead of two. If the type of rod you are using has a fully floating pin this is fine, but if you intend to modify the stock rod to make it fully floating then it would be best to read more on the subject later.

If you choose to use a press-fit pin, then you need not concern yourself any further with pin retention. But if the rod is to be fully floating, then how the pin will be retained in the rod must be considered. When buying pistons, do not assume that all forged pistons have pin retention that will take care of either fit or floating assemblies. Some forged pistons do not have the circlip grooves necessary to use a floating pin and machining in a circlip groove can be an expensive operation.

As of 1990, the trend is to machine pistons so they can be press fit or floating. Due to piston-to-crank throw centerline offset, slop in the piston, clearance in the bearings, flex of the crank and other factors, pins can hammer against the pin retainers, causing heavy side loads. For this reason double circlips or Tru-Arcs (often termed double locks) are advised, which is what most piston manufacturers use these days.

Double locks involve the installation of two circlips in each groove. Before installing these circlips, take a close look at them. Essentially, in their manufacture they are stamped out of high-carbon sheet steel. Inspection shows that one edge of the circlip is slightly rounded, whereas the other side has a sharp edge. When installing the circlips, butt the two rounded sides together so the flat edges are against the pin and the outside of the groove.

Also popular with many engine builders are Spirolocs. As the name suggests, these are flat-wound spiral retainers that are installed in the pin retainer grooves by threading-like action. Many engine builders have a love-hate relationship with Spirolocs. They love the reliability or security they give, but hate to install or disassemble them.

If a full floating rod is to be used, the pin must be retained in the piston by some means other than the now-defunct press-fit in the rod. The most common technique used for volume-production high-performance pistons is the double circlip at each end of the pin.

Custom Aftermarket Pistons

5

Presuming that all the big piston manufacturers know what they're doing when it comes to making pistons, why would anyone with less knowledge of piston manufacture want to trust themselves to lay down a piston specification for their own motor? Good question! The custom piston industry exists precisely because many engine builders do want to specify certain parameters for their own pistons. It isn't that the big piston manufacturers don't know what they're doing; on the contrary, the majority are skilled at their art. But the pistons they manufacture are to suit as broad a range of applications as possible, and all specifications are on the conservative side. By laying down the specification of a custom piston, the engine builder is attempting to get a piston that most closely suits the requirements.

Why Choose a Custom Piston?

So where does a mass-produced piston fall short compared to the custom piston? Remember that a mass-produced piston must be competitively priced for today's market. It also has to be sold through wholesalers and jobbers before it gets to you, the customer. So there are prob-ably two markups on it, sometimes three. Thus the production costs must be kept to a minimum.

With a custom piston you will likely deal directly with the manufacturer, and you'll be paying its price. Though the manufacturing costs of a custom piston are higher, generally speaking there is less markup.

Because a mass-produced piston must be reliable, quiet and reasonably priced, compromises must be struck. Consider, for instance, piston ring position. A high top ring is better for power because there is less space in the ring land between the piston and

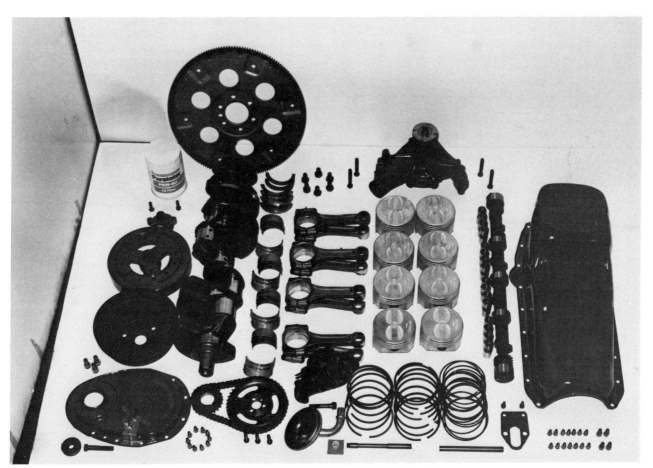

A lot of parts go into the bottom end of a small-block Chevy, but certainly amongst the most important to get right, whatever the application, is the piston and ring assembly.

cylinder wall where dead fuel and air can accumulate, which rarely plays a part in the combustion process. For an off-the-shelf production piston, the position of the top ring must be determined based on a worst-possible-case scenario. For instance, the piston could be used in an engine that is run too lean and with nitrous-oxide injection; the manufacturer of a mass-produced piston must put the ring down the piston far enough to cope with the high heat developed by such a situation.

On the other hand, anyone capable of designing a custom piston should know the application in the first place and therefore set up carburetion and timing appropriately. Let's say an experienced engine builder is putting together an engine for circle-track racing with its fueling, compression and thermal characteristics as known quantities based on experience. Thus, the engine builder can place the top piston ring in a more suitable position.

Knowledge of and experience with pistons can be beneficial in other aspects as well. For instance, an engine builder can work with a custom piston maker to keep engine weight to a minimum. Reciprocating weight in a piston costs power; excessive weight beyond what is necessary to build a strong enough piston for the job causes increased cylinder wall and bearing friction. Heavy pistons also mean a greater amount of counterweight on the crank. The amounts we are dealing with may be small, but if taken care of will lead to a more powerful engine.

Custom Piston Makers

There are many companies producing quality pistons, including Arias, Cosworth, J. E. Ross and Wiseco. Any one of these companies will more than likely be able to produce a piston to your requirements. But before choosing, there are a few questions you should ask.

First, find out what material your piston is likely to be made of, and what cold clearance is likely to be needed. Will these suit your intended application?

Second, discuss your piston requirements with several companies; one company may specialize in an application that fits your needs and have a more or less off-the-shelf piston available. This makes it less expensive than a custom piston from another manufacturer that doesn't specialize in that style of piston.

Finally, ask what the delivery time is expected to be. Some of these piston manufacturers, by virtue of reputation, get backed up at the beginning of the race season, and if

Starting from a heat-treated, forged blank, here are the various main stages a Ross piston goes through to achieve the end result.

Note the position of the ring pack on this TRW piston. The top ring is positioned well down from the deck of the piston to shield the ring from excessive heat and to ensure the integrity of the top ring land.

This custom-machined piston has the ring pack much nearer the crown, and is intended for use in an all-out drag-race engine.

your intention is to go racing, your pistons may not arrive until mid-season. The clue here is to start the ball rolling in November and December.

Once you've decided on a piston manufacturer there are certain factors you will have to consider to ensure that whatever company makes your pistons can produce exactly what's needed.

Piston Manufacturing Information

To make a custom piston, the piston manufacturer will need to know the type and specification of the cylinder head you intend to use. Details will include whether or not it is angle milled, whether it is being used with offset block dowels, and whether the valve centers differ from stock as is often the case in aftermarket heads.

Additionally, the piston designer will need to know the compression ratio you plan to use. To make a compatible piston dome, the designer will also need to know the displacement of the combustion chamber and the amount of deck clearance. Ideally, deck clearance should be as small as possible without the piston hitting the cylinder head at maximum rpm.

After the compression ratio has been dealt with, you'll need to decide on ring width and their position on the piston. For a high-output engine, another important factor is the rod length that will be used; together with the stroke length, this will control the piston compression height. For what it's worth, it is difficult to get the compression height to anything less than about 1in and over 1.8in the pistons start to get needlessly heavy. To get any given combination into a stock small-block Chevy the rod length, plus half the stroke, plus the compression height must not exceed the 9.025in crank centerline to deck height.

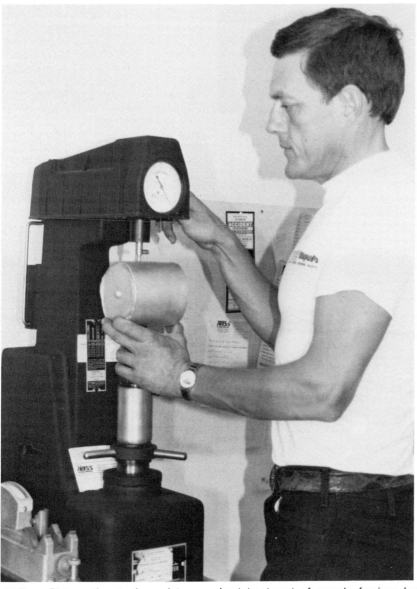

At Ross Pistons, the starting point, as with many others, is from a forged, heat-treated blank. Here, Moe Mills checks the heat treatment of sample forgings by hardness testing.

After boring from the underside so that they can be mounted on a spigot, the blanks are located on this machine. It produces the ring groove pack and machines the OD to rough turned size ready for finishing.

Next, you need to consider wrist pin selection. Factors to be decided on include pin length, wall thickness, diameter and retention method.

Additional thought must be given to the oiling system for the pin. Often, to be able to shape the lower edge of the piston skirt, it's necessary to know what crank will be used in the engine, as the counterweight clearance must be taken into account.

Of course, application is the primary factor to consider, here. Will the pistons be for a blown motor? Are heavy loads of nitrous contemplated? Will it be an alcohol or nitro burner?

For the most part, the piston manufacturer will ask what you require, rather than give advice on what they think you should have. This puts the onus on you as to whether the piston works adequately. Of course, all of these piston manufacturers know full well what usually works and what doesn't, and will give advice when asked. Still, you'll need to be certain of what's needed.

Designing Your Own Pistons

One of the first considerations in piston design is what the engine can tolerate in cold clearance. This will influence both the alloy and the design of piston that should be used.

In the previous chapter we touched on the subject of cold clearances; now we'll take a more in-depth look. One of the aspects we seek to control by keeping piston clearance to a minimum is the stability of the rings on the bore. If the piston rocks too much in the bore, it changes the angle the ring face makes with the bore allowing blowby leakage.

The lowest blowby that I've recorded has been with pistons running as little as 0.002in cold clearance. The pistons were hypereutectic and used in an engine with 320hp. With a conventional moly top ring, taper-faced second and 0.1875in (3⁄16in) oil control ring set, zero

Some piston skirts are ground, some are turned. Whichever way, the skirt must have a special form on it to compensate for expansion. These particular pistons are being ground to produce the desired ovality and barrelling.

This special machine at Ross Pistons puts the valve cutouts in the piston crown.

Here a purpose-built machine is being used to form the piston crown on the spark plug side of a small-block Chevy piston. The template on the left side of the table produces the contour on the piston just off center right of the picture.

blowby was measured using a Dwyer blowby gauge. Normally, blowby figures are in the range of 2-3cfm for a good forged piston with conventional rings.

If you're trying to build a motor for maximum power for whatever application, be it a street or Winston Cup racer, then you must consider how close the piston approaches the cylinder head quench area. This dimension is controlled by the piston-deck height. As a piston goes over TDC (top dead center), it rocks; the more clearance the piston has, the greater the amount of rock. When the piston rocks it uses up some quench area clearance, meaning that a looser-fitting piston requires more quench area clearance than a tight-fitting one, and therefore the effectiveness of the quench will be reduced.

There's another way to view the situation. If you're using a high-expansion piston and keeping quench area clearance to a minimum, you'll need to thoroughly warm the engine so the piston is expanded prior to turning any significant rpm. This should be standard practice for a race engine anyway. If it is a street motor, you'll need to keep the clearances to a minimum to reduce or

Pistons can come in a great variety of weights. Many pistons are designed to be a direct replacement for the stock weight piston, so, of necessity, will weigh in at around 550g. Other pistons are made just as light as is practical. From the range of pistons here, it can be seen a difference of 200g exists between the heaviest and lightest pistons. The lightest pistons tend to be those used with longer rods, for instance the flat top Wiseco piston for a 6.00in rod, weighs only 387g. The 426g piston next to it is a Ross piston for use with a 5.70in rod; this is among
the lightest 350 pistons for this rod I've used. The additional weight on this piston over the lighter ones utilizing a 6.00in rod is due to the fact that there has to be 0.030in extra compression height to compensate for the shorter rod. Obviously a longer rod means a lighter piston, but on the other side of the coin, more rod means more weight. The question is, does the reduced weight offset the increased rod? With a steel rod, it's marginal. With a carefully considered choice of rod and piston, the little end weight combination can be slightly lighter with
the longer rod and the shorter piston. If an aluminum rod is used, then a comparison with a stock steel rod and height of piston shows the longer aluminum rod and shorter piston to be far lighter. The same goes for titanium rods. However, if we're comparing a 5.70in rod with a 6.00in rod in any given material, the weight change of the reciprocating mass at the pin end is only marginally reduced; however, the long rod makes for a better rod-stroke combination.

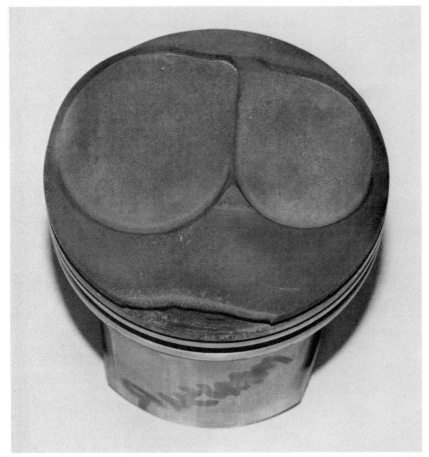

This piston came from an alcohol engine that was run too lean. See how the ring land has started to erode and peel up due to excessive heat and inertia loads over TDC.

eliminate piston slap during warm-up.

The amount of cold clearance involved depends both on piston material and design. Pistons are usually made in two basic forms: the full-circle piston, and the slipper piston. Oddly enough, the full-circle piston is not perfectly round, rather it's oval-shaped so that only the faces on the thrust axis touch the cylinder walls. From this, we could conclude that the sides of the piston at the pin bosses are redundant.

In fact, this is the thinking behind a slipper piston. It has no sides contacting or even near the cylinder bore and would seem an obvious way to go. Unfortunately, life is rarely this simple; many factors complicate the issue with expansion and how to deal with it topping the list.

Piston Skirts

In designing the skirt you must consider what happens to the piston form when heated. With a full-circle piston the expansion tends to be more uniform because the thrust face expansion is influenced by the full-form skirt. On the other hand, a slipper piston has two walls passing all the way across the piston much nearer to the center. When these walls pick up heat from the crown they transfer the expansion directly across the skirt. Consequently, getting the slipper piston's skirt right is more difficult but not impossible.

The success of the form on the piston's skirt can be measured by how much the piston wears the cylinder bores. Cylinder bores that wear rapidly in the areas in which the skirt contacts the bore are indication that the skirt form is not perfect. Generally speaking, full-circle pistons tend to be easier on bores. However, on the negative side, full-circle pistons tend to be heavier than slipper pistons. So it is necessary to consider where your priorities lie.

Piston Crowns

The next issue to contend with is the piston crown. This area can be so critical to a high-output engine that sometimes engine builders buy their pistons with the crown unfinished, then custom fit the piston crown to the application. Their intent is to keep the crown height to a minimum, and this often means using valve cutouts of minimum depth consistent with the cam being used. These ul-

Mike Parry, one of England's brightest rising race engine builders, checks blowby as a means to relate to prevailing ring gaps under load.

tra-critical domes are more common on smaller engines because you need a larger dome to create a high compression ratio in the smaller combustion chamber. If you are going to produce your own piston domes, then some handwork will be involved. In addition, it will be necessary to machine the valve pockets, in which case you'll need a good piston vise.

By machining their own piston domes, engine builders attempt to minimize the effect of tolerance stack-up. In other words, the piston domes can be precise for the existing parts combination. Due to the lack of machining facilities, however, most will obtain custom pistons with the crowns finished. This being the case, there are certain aspects to be aware of.

First and foremost, for high-compression use the cylinder heads should be milled to the minimum chamber volume possible. The reason for this is to minimize piston dome height; the higher the dome is, the harder it is for the flame front to travel effectively through the charge. The chamber shape begins to be compromised when the dome height gets higher than about 0.150in. To keep the dome height to a minimum, the chamber must be as small as possible and the piston needs to be as near zero deck as possible.

If you know the rpm of the engine is not going to be too high, and that the crankshaft and rods being used are adequately stiff, it is possible to bring the piston out of the top of the block a little ways to cut the piston-to-quench-area approach distance. Of course with a bigger engine, the dome becomes less of a problem. Big engines can generate high compressions with bigger chamber volumes. For a 406ci engine, a 13.0:1 compression ratio doesn't represent much of a dome.

There are situations where the combustion chamber needs to be as large as possible. Say, for instance, you are building a supercharged, large-displacement small-block, and you're looking for a compression ratio in the mid-eights. Under these circumstances the piston could end up with a considerable dish in it, but dishes past about 21–22cc are going to be a full circle, so much of the piston quench area is lost. On a supercharged engine, quench can be important, both for part-throttle fuel economy and for suppressing detonation.

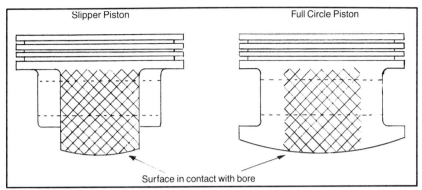

Here is the basic difference between a slipper piston and a full circle piston. Though the slipper piston is lighter, it tends to be more difficult to control expansion on the skirt. This means the skirt profile is more critical. Often a full circle piston, as seen on the right, will be easier on bores.

An underside view of these two pistons easily distinguishes the slipper type from the full circle one.

These pistons typify the two main styles of valve cutouts: the trough and the eyebrow cutouts. For a given compression ratio, a piston employing the trough cutout has to use a higher dome, which may, if taken too far, inhibit the flame front.

Piston Coatings

The two major problems with aluminum pistons are that they conduct heat rapidly, and the strength drops as temperatures rise. Aluminum pistons could be improved if the heat conductivity and operating temperature could be reduced. These two factors can go hand in hand to deliver improvements.

If the combustion chamber heat can be more effectively contained instead of soaking through the piston, two advantages will be realized. First, the piston will be stronger because its temperature is lower, and second, the engine will develop more power because more heat is retained in the charge, thus raising cylinder pressures.

The idea of insulating the piston crown from the rest of the piston is not new. Attempts were made even prior to World War II, but unfortunately the technology of the era was not up to the task and most early efforts resulted in this usually abrasive coating peeling off, causing engine damage. Technology in the aerospace and allied industries made advances in coating techniques in the late 1960s and early 1970s.

Today there are a variety of functional coatings that can make *small* gains in engine performance and *significant* improvements in piston reliability. This can be especially important for a long-distance engine where leaner carburetor settings for reduced fuel consumption and thus fewer pit stops can mean running closer to detonation limits.

Banodizing

I've successfully used coatings from three companies. The first of these, a seemingly unique process known as Banodizing, is produced by Lovett Industries in Santa Fe Springs, California. This process appears more akin to a plating process than a coating application as the basic procedure involves an electrochemical process carried out in a vat. Banodizing produces part build-up and part penetration. Typically, if the coating is 0.002in thick, the surface will grow by about 0.001in. Therefore, the process can make a standard piston up to 0.002in bigger in diameter.

Even though its resultant coating is thin, this process produces a surface that acts as an effective heat barrier, making piston strength greater by virtue of reduced operating temperature.

Banodizing also improves the fatigue life of an aluminum component. If fatigue failures were becoming a problem, Banodizing could well add 50 percent or more to the life of the piston. Additionally, the coating usefully reduces the friction coefficient of aluminum.

A third benefit is that the Banodized finish is much harder than the base aluminum and in fact, its hardness is comparable to steel. If suitable machining margins are allowed, the entire piston can be treated, including its ring land area. After Banodizing, the ring lands are far less likely to pound out or show any significant signs of wear, even after extended periods.

Virtually all types of pistons can be Banodized, but some piston alloys are more suitable than others. The least effective materials for Banodizing are high-silicon alloys. Generally, piston brands such as Wiseco, TRW and Sealed Power will build only 0.00075in of Banodizing, whereas the low-silicon pistons such as Cosworth, Ross and so on can be Banodized to a depth of about 0.002in.

Though I've applied the coating only to a limited number of engines, my experience with Banodized pistons has been favorable. The process is inexpensive, and the turnaround time for sending pistons to Lovett Industries is short.

Swain Industries Coatings

A company that has done much pioneer work on coatings is Swain Industries, headed by Dan Swain. Swain Industries specializes in many types of coatings for various engine components. Early types of Swain coatings used a sprayed-on zirconium-oxide that produced about a 0.015in thick insulating layer on the piston crown.

During the 1970s this was the most common technique used. I've tested engines with pistons coated in this fashion, and the most immediate difference that I noted was a substantial reduction in heat going to the oil. Another factor was that when test engines were run with detonation for a number of tests, I found no piston damage where at least some was expected had the piston been uncoated. Also, under-crown piston temperatures never reached that of uncoated pistons, as evidenced by the lack of temperature-induced oil stain on the underside of the crown.

The current process from Swain Industries for insulating piston crowns is entirely different from these early methods. Again, I've performed lim-

By the late 1980s and early 1990s, typical piston coatings have changed much in character since their use started some dozen years earlier.

Many coatings, such as this Swain treated piston here, have molecularly bonded coatings that cannot detach themselves as could earlier coatings.

ited but wholly successful tests on these pistons. The current coating involves molecularly bonding a coating of only 0.002-0.003in thickness. To reduce skirt friction, Swain offers antifriction coatings, usually Teflon based. These coatings are applied to piston skirts with the idea of reducing piston-to-cylinder-wall friction; such coatings can also be applied to crankshafts and connecting rods. When applied to these components the oil has less tendency to stick to the finish, and therefore it sheds faster.

Polydyne Coatings

The third coating company that I am well acquainted with is Polydyne Coatings of Texas. Polydyne coatings were tested some years ago on a limited number of engines. At the time, the zirconium-oxide coatings were in vogue and Polydyne was used on the piston crowns.

Essentially coatings look like a good idea, but do they deliver? They do provide greater piston reliability, reduced oil heat and a measure of protection against detonation. As far as horsepower, various claims have been made that increases up to 10hp on a Winston Cup engine are possible. Although I've never seen results of a true back-to-back test, personal experience indicates that gains of 4-6hp are a reality on a high-output small-

block Chevy engine that has been coated. Though these gains in power are small, they are welcome.

The biggest asset of thermally coated and Banodized pistons has been the

considerable extension of piston life. Coated pistons have often been reused in as many as four engines, each of which has been on the dyno for as much as one month's testing.

Teflon coatings on piston skirts are also becoming popular. The idea behind such coatings is that the sur- faces hold more liquid and they'll slide over the oil film easier.

If you're going to build a 400ci motor with an 8.5:1 compression, for example, a 21cc dish on a piston that comes flush to the block is going to need a large chamber in the range of 80cc. A 21cc dish is about as large as can be made in most pistons consistent with the cylinder-head chamber form and quench area, although there are some hypereutectic pistons for the 400 block that are about 30 cc. The mirror-image piston is often referred to as a reverse-dome piston.

Valve Cutouts

Custom pistons have definite advantages over many off-the-shelf pistons in creating valve cutouts. Most off-the-shelf pistons employ a simple trough for valve clearance, but the trough valve cutout has two disadvantages.

First, to achieve a given compression ratio a higher piston dome is needed to compensate for the wasted space between the two valves. Since

the cutouts do not conform closely to the shape of the valves there is material missing from the piston that could add to the compression and reduce the dome height required.

Second, if the valve cutouts follow the form of the piston, then around the TDC mark the valve cutout tends to shroud the intake from the exhaust. As a result, there is less scavenging of fresh intake charge through the exhaust. All other things being equal, the eyebrow valve pocket-type piston can generate the same compression with less dome and reduce the possibility of scavenging fresh intake charge out the exhaust.

Though apparently not critical, the form of the cutout can have influence on the engine's power output. It is desirable to shroud the intake and exhaust valve where their circumferences are close together to reduce the likelihood of crossflowing, but it is not necessarily desirable to shroud

them around their entire circumference. For this reason valve cutout centers can often be pulled farther across the piston than the actual centerline of the valve. Also, it is necessary to have a reasonable radius at the bottom of the valve cutout; when you get the piston back from the machine shop, hand finishing of the top corner of the cutout is needed so it ultimately generates the proper form.

On the surface, the semi-circular valve cutouts appear to have no drawbacks. However, sharp-cornered cutouts with a pointed section where the two join can raise thermal stress. Also, the thin section of the piston where the two cutouts meet can get hotter than the rest of the piston since it is near the center of the piston with a lot of exhaust flow. Fortunately, some cooling takes place due to the intake charge.

To minimize any negative effects, the cutouts should have a radius at

For engine builders who desire to machine their own piston crowns, I recommend the use of this Impulse Engineering piston vice. In 1990, I switched over from the more laborious methods of holding pistons to this vice. With the change came improved accuracy and reduced machining time.

both the top and bottom. Additionally, the beak formed where the two cutouts join should be formed so as not to leave any sections less than about 0.090in. Thus, an eyebrow-type piston can have a slightly stronger crown than a trough cutout.

Achieving this situation is not always as simple as it may sound. Often a piston crown which makes numerous changes in section thickness may oil can where the piston crown suddenly deflects a relatively large amount for little increase in load or temperature, much like the bottom of old-fashioned oil cans. If this becomes excessive it can lead to fatigue at and around the beak and base of the valve cutouts.

Now you need to consider the cutout depth. With a single-pattern camshaft, the cam will almost always be advanced in the engine by about 4deg. This means the intake valve will always make a closer approach to the piston than the exhaust and therefore the intake valve cutout will need to be deeper.

For a dual-pattern cam, the exhaust duration is often longer by 8-12deg. This usually means the valve cutout for the exhaust will be of similar depth to the intake and, depending upon the rocker ratio, sometimes deeper.

All of these clearance requirements can be overridden by the need, on a high-rpm engine, to take into account

the effect of valve float and bounce. The intake closure takes place when the piston is a considerable distance from the valve, but this is not the case for the exhaust. Valve closure for the exhaust takes place while the piston is at the top of the bore. If valve float occurs, it is most likely the exhaust

Getting the compression low enough on big-inch motors can be a problem. These special Brodix heads, available through Whipple Industries and C&G Porting, not only sport the necessary oversize exhaust valve at 1.80in, but also have chamber volumes up to 82cc. This will allow compression ratios as low as 8.0:1 to be achieved even on big-inch engines.

Heat Treatment

An important part of piston manufacturing is the heat treatment of the parent material. Although precise heat treatment applied to pistons may vary from manufacturer to manufacturer, for the most part the processes can be described as a T4 solution heat treatment followed by a T6 or T61 precipitation heat treatment.

Basically, the forgings are heated to a temperature near the point at which the material becomes plastic. This is typically around 950deg. Fahrenheit. The pistons are held at this temperature for an extended period of time, which, in the case of Cosworth, runs twenty hours.

What the solution heat treatment does is put the alloying elements into solution with the aluminum rather than have them form at the grain boundaries. After the appropriate time at the elevated temperature, the piston forgings are quenched in boiling water. By rapidly dropping their temperature, the alloying elements don't have time to come out of solution and form at the grain boundaries.

After quenching, an aging treatment is performed. This involves heating the pistons to a temperature between 350-390deg. Fahrenheit for between ten to twenty hours. This process artificially ages the piston to stabilize the hardness. The final heat treatment process is similar to the process pistons are subjected to in operation.

As expected, much of the heat treatment effect applied to the crown is lost. The high temperatures seen by the center of the piston will actually anneal the piston at this point. However, the heat treatment is largely retained at the ring belt and wholly at the skirt and pin areas where it is the most important.

The fact that piston hardness can change due to local temperatures can be used as an indicator of cylinder temperatures, and in turn, mixture distribution. By hardness checking the pistons across the crown and referring to a hardness versus temperature scale, it is possible to get an idea of the working temperatures applied to various parts of the piston.

that will contact the piston. For this reason it is better to have either additional clearance for the valve cutout, or, if maximum compression is a prime requirement, stiffer exhaust valve springs. The extra spring force will ensure the intake float first and thus act as a simple rev. limiter.

Valve Cutout Depth

The question is, how much clearance should be allowed between the valve and piston at closest approach? The first thing to remember is that the closest approach is not at TDC; with a single-pattern cam the intake is usually closest to the piston at about 10-20deg. ATDC (after top dead center). With this in mind, it may appear as if installing custom pistons is too much trouble since you must measure piston-to-valve clearance prior to ordering them. This entails making a dummy build of the engine. Use clay on an old piston to determine exactly what clearances exist, then measure this and turn the dummy over to the piston builder.

The majority of piston manufacturers, however, have a pretty good idea what depth of valve cutout is required for each cam type used. Remember, it is unlikely that your combination of cam and cylinder head will be totally unique. As a result, if the piston manufacturer is given the basic specifications of the cam, the amount of deck clearance (plus or minus) and the valve lift at split overlap, it can come up with a valve cutout depth close to the mark.

Cam and piston manufacturers usually shoot for 0.100in clearance between the valve and piston when the engine is turned over by hand. Of course, in operation these clearances could decrease, as crankshaft flex, rod stretch and so on allow the piston to move farther up the bore. With vibrations in the valvetrain, valve float and so on, the valve could be farther open at the point of close approach than predicted. The net result is that some of the clearance could be used up. But if the characteristics of the combined parts are known, valve-to-piston clearances can be reduced significantly.

Some engines have managed to squeak by with as little as 0.030in static clearance. Yet, one missed shift or one trip into valve float could mean sixteen bent valves. Usually, 0.060in clearance is safe in a known engine. Nevertheless, if you are building an engine for the first time and you don't know how much those particular rods stretch or how much the crank flexes and so forth, then aiming for 0.100in clearance is the safest route.

What about the dome in a small-cubic-inch engine that must have a high compression ratio? The problem is to get the flame front to climb over the dome and burn the main charge mass.

Success hinges on starting a good initial burn around the spark plug so that it has enough force to light up the

Though still requiring some hand finishing, this piston crown typifies the preferred form. Much of the eyebrow is still retained, but thin sections at the beak have been removed to avoid thermal stress raisers.

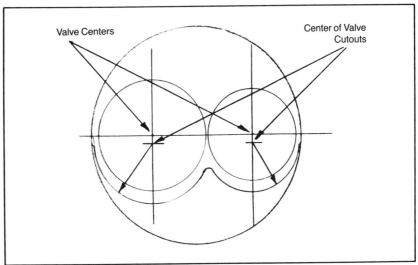

In this plan view of the piston, note how the valve centers don't actually coincide with the centers of the cutout. This allows the cutout to cause less shrouding of the valve during the time when cylinder scavenging is in operation at TDC overlap.

mixture over the dome. Therefore, it is a good idea to machine the crown so there is a good amount of charge in the vicinity of the spark plug and so that the electrodes are not shielded by the dome.

Since ignition takes place about 40deg. BTDC (before top dead center), much of the combustion occurs when the piston is still compressing the charge. As a result, the manner in which the burn takes place during this phase is important. When the plug lights the mixture there is an ignition delay. A small kernel of flame is generated at the spark plug but does not expand rapidly for the first 10-15deg. When it begins to expand it is still slow but by the time the engine reaches TDC, the expansion takes on a meaningful burn rate. Consequently, the most significant part of the charge is burned during the piston's first ⅛in travel down the bore. Another way of looking at it is that 80 percent of the power the engine generates takes place during the first 20 percent of the stroke. Thus the way the flame burns across the piston is important.

To get this flame going it is helpful to use a small dish that extends to a fire slot across the crown. At first this may seem like an effective way of reducing compression; however, the need to obtain good burn is far more important than the need to get the ultimate compression. Losing a half a point of compression may only cost the engine 10hp. Yet allowing the chamber to burn badly because of the crown shape can wipe out 150hp.

In general, the lower the spark plug is in the cylinder head, the more critical the crown form is. Most modern heads used for high-performance and race work have the spark plug situated as near to the roof of the chamber as possible, allowing the flame front to pass as easily as possible over the crown.

On the other hand, on many production heads the spark plug is located relatively close to the cylinder-head deck face. These cylinder heads are much more critical in terms of dome shape. For these pistons, it is important that the tops of the dome are rounded off. With many pistons, much of this rounding off will need to be done by hand. Indeed, it's a good idea to hand finish all domes, since most machining operations leave sharp corners. A needle file and a Scotch-Brite pad for smoothing work fine. The piston may not look as shiny and flashy when you are done, but it will work better.

When the cylinder head has a high spark plug location immediately adjacent to the roof of the chamber, the flame slot across the piston crown is not as critical. The slot in this piston typifies what many engine builders use, though I like to see it just a little larger than this.

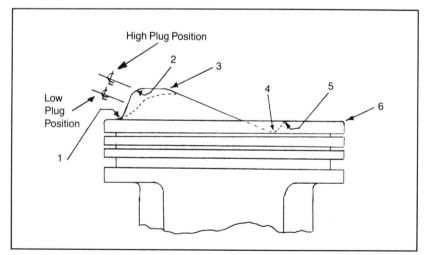

When ultra high compressions are used, especially in small-displacement engines, the dome shape on the piston can be influential. Higher domes can be used with a high plug position common to high output aftermarket and Bow-Tie heads. However, with the low plug position of many stock heads, it's necessary to cut the piston away to the dotted line to retain an effective flame front propogation.

Piston Crown Thickness

Much of the material in a piston resides in the crown, ring belt and pin pillars. It is difficult to lose any significant piston weight by reducing the skirt, although every little bit helps. Yet if the piston crown could be made thinner, a substantial weight savings can be made.

The question of how thin the crown can be depends on many factors, but the prime considerations seem to be the material the piston is made of and how much oil cooling takes place on the underside of the piston. The actual physical design of the piston can play an important role. Obviously, a piston with a large, unsupported area in the middle of the crown is more likely to fail than one having support toward its center.

Typical crown thicknesses for a Chevrolet piston range between 0.320-0.400in, but there are pistons that fall outside this limit. For instance, some hypereutectic pistons have crowns as thick as 0.550in. This is not necessarily because the material is needed, but because the piston is intended as a stock replacement and so must weigh the same as the stock piston. Stock cast pistons are heavy, since in most cases they contain a steel strut. In the replacement piston the weight had to go somewhere, so the manufacturer chose to put extra material into the crown, thereby making a relatively bulletproof piston.

Where the valve pockets are located, a thinner crown thickness can be used. Experience indicates that with the piston types commonly used, crown thicknesses shouldn't drop below about 0.150in at the thinnest point and have at least 0.200in over most of the surface area.

Clever piston design can pay off with thin crown thicknesses. For instance, Cosworth produced a big-block Pontiac Pro-Stock piston that had a crown only 0.125in thick in the center. This isn't bad for a piston of such large diameter and the abusive circumstances in which it runs. Typical failures occur through the valve notches with the crack spreading from the center of the piston and eventually reaching the pin bore. This alone should demonstrate why we need to have a generous radius in the valve notch.

One of the problems with pistons is that they are made from aluminum. This may be a good material to use as far as weight is concerned, but its strength drops off quickly with rising temperatures. Anything that can aid the cooling of the piston crown helps

the piston to retain its strength. Oil splashing on the piston's underside acts as a coolant, yet for most high-rpm, normally aspirated race engines we go to a lot of trouble to make sure there is minimal oil flying around in the crankcase. It costs power for the rotating parts to cut through the oil, thus too much oil in the crankcase causes windage losses. Therefore, cooling the piston via oil is not the way to go for such an engine.

We can look at the problem the other way around, however, and prevent heat from getting to the piston in the first place. An acceptable tech-

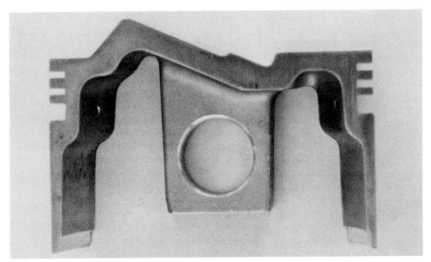

This Cosworth piston typifies the sort of cross-section thickness to produce a relatively light yet sturdy endurance-racing piston.

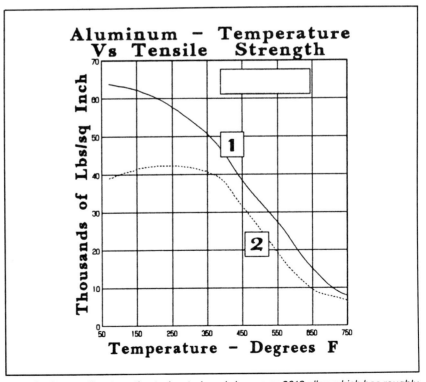

Here is the tensile strength at elevated temperatures for Cosworth's RR58 alloy (1). Many of the other piston manufacturers utilizing a low-silicon alloy use what is known as 2618 alloy, which has roughly similar properties. The dotted line (2) shows an alloy typically used for many OE-type applications.

nique for this is to coat the crown with a thermal barrier, reducing the heat passed into the crown. The net result is that the crown thickness can be reduced.

On a supercharged engine the temperature of the piston can be significantly higher than on its normally aspirated counterpart. In fact, when an engine is supercharged or turbocharged at high pressures it may be necessary to direct an oil cooling jet to the underside of the piston simply to help the piston function irrespective of the crown thickness. The amount of windage losses may be high, but the return in increased boost is higher.

Since the circumstances in which a piston operates in a supercharged engine are far more severe, it is imperative to look at all the ways possible to upgrade the piston's potential strength and reliability. If thermal coating works well on a normally aspirated engine, then it's certainly good for a supercharged engine.

There are some thermal barriers that are applied in a similar technique to anodizing, in particular Banodizing. This is a coating that has proved effective in both normally aspirated and supercharged test engines. It is important to understand that this coating should be applied to the outer surface of the piston *only*. If applied to the underside of the piston it will prevent the piston from releasing existing heat, and cooling from the oil will be less effective. What is needed is a barrier on the outside and a plain aluminum surface on the inside.

Ring Pack

The only reason for building a custom piston is for more reliable power output. One of the principle design areas that can aid power output is the ring pack.

How the ring-pack components are sized can have a significant influence on the power output of the engine. For instance, thinner rings can allow a more compact ring pack, which in turn will allow for a longer connecting rod, something that large-displacement small-block Chevys need.

The position of the rings down from the deck face of the piston can also affect rod length and how well the piston allows combustion to take place. Gases trapped between the top ring land and the bore don't aid in the combustion process; if and when they do burn, it is too late in the power stroke to have any effect. If the ring is too close to the top of the piston, the ring land becomes too thin and flimsy to support the inertia loads. If it collapses or transmits too much heat to the top ring, it will prevent it from doing its job.

Just how far down the piston can we make the piston ring groove? This depends on the application. Bill "Grumpy" Jenkins, one of the world's leading small-block engine builders, has used as little as 0.060in for a drag-race engine and got away with it—although he claims 0.080in is a better figure. For drag racing this may be fine; however, you better be sure the mixture never runs lean as it would take only a brief moment of detonation to close up that ring gap.

Unless you are extremely sure of your settings, even for drag racing I recommend using a top ring positioned at least 0.100in down from the deck. For an endurance-race engine, figure 0.150in down from the top; again, carburetor and ignition calibrations must be right on.

If you're unsure of engine settings, it would pay to have the rings 0.025-0.050in farther down. On most production pistons the rings tend to be between 0.300-0.400in down. In so placing the ring the piston manufacturer is attempting to cover themselves, as far as possible, from the more common malfunctions that may occur. This is not necessarily where it wants them to be, but this is where they are safe.

Another factor that can dramatically affect the final position of the top ring is the depth of the valve cutouts. If valve pockets are deep, the valve pocket edge will intersect with the ring groove. Often this is the thinnest part of the crown, but because it is supported by the ring the required thickness here can be minimal. Some pistons use as little as 0.030in.

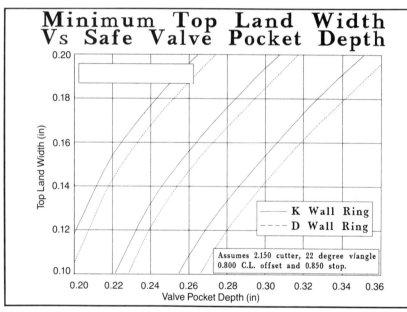

This chart shows how close a ring groove can approach to the top of the piston without intersecting the valve pocket. The two curves on the left of the graph apply to 3.750in bore engines. The middle pair refer to 3.875in bore engines, and the right hand pair are for 4.000in bore engines. The solid and dotted lines of each pair refer to the different depths needed depending whether a K wall or a D wall ring is used. The K wall ring has a radial depth equal to the cylinder bore divided by 20. The D wall has slightly less radial depth and can be determined by dividing the bore size by 22. If a Dykes top ring is used, then for bores between 3.810in and 4.500in radial depth will be 0.175in. If you're looking to get the top ring as close to the top of the piston as possible, then a Dykes ring, because of its usually lesser radial depth, can be brought nearer the top of the piston. The method of measuring the valve pocket depth used for this graph is the vertical distance from the deck face of the piston to the corner of the valve pocket.

Ring Land Volume

The next question is, just how much can it be worth to save, at the most, 1cc of ring land volume by moving the ring up to as near the crown as possible?

Changing the volume by only 1cc in a 350ci engine containing 747cc per cylinder seems insignificant. However, the situation should not be taken at face value. When the engine inhales a charge and fills the cylinder, only 1cc of charge will reside in the ring land when the piston is at the bottom of the stroke. As the piston compresses the charge, the pressure rises and the amount of volume contained in the ring land becomes a greater proportion of the amount above the piston. By the time the pistons arrive at TDC, what was originally 1cc of volume could be equal to as much as 10-14cc at normal pressures. This 14cc out of 747cc is a noticeable percentage of charge now residing in the ring land. Yet the situation gets even more complicated.

When the spark ignites the charge, the flame front begins and pressure build-up follows. As the pressure continues to build in the chamber, it drives more of the remaining unburned charge into the ring land, an area likely to be the last part of the charge the flame front will reach.

To make the most of the ring pack it is important to understand how it functions. The goal of the compression ring here is not to seal by virtue of its own radial pressure, but to be assisted by the gas pressure generated above the piston. For this sealing to work it is essential that the gas pressure get behind the piston ring to force it onto the bore. In a conventional ring groove it must have a certain amount of top-to-bottom clearance. Also, the ring must seal against the bottom of the groove, as any gases passing under here will relieve the radial pressure on the ring and cut its effectiveness.

The question is, how much top-to-bottom clearance should there be for maximum effectiveness? If the distance is too much, the ring will not be contained in the groove properly and could, over TDC, slam against the top of the groove. As pressure is applied it will push the ring down into the groove, yet not sufficiently for it to seal against the bottom and gases could escape around the ring. This ring flutter can cause serious power

Ring Tension

Since it can measurably affect power, ring tension is an important parameter for the high-performance engine builder. At this time, there is no clear standard within the industry as to how to measure ring tension.

A reasonably common technique, known as diametral load measurement, is to load the ring at 90deg. to the gap and compress it down to the cylinder bore diameter. Although this technique produces a repeatable figure, it does not actually represent the radial force the ring exerts on the bore.

A more realistic procedure is the *tangentral* load measurement, where the ring is loaded at the gap and compressed to the bore diameter. This technique more accurately measures the radial force the ring exerts, but it does assume that the force is uniform while, in practice, this often is not the case.

Depending upon the consistency of manufacture, the pressure pattern the ring may exert in the bore can vary considerably. Making rings more consistent in their radial loading is the job of the piston ring manufacturer, not the end user.

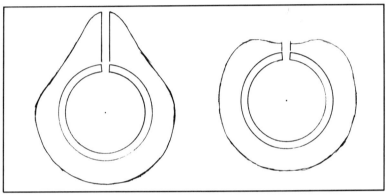

Cylinder wall loading patterns, delivered by a particular type of ring, can vary considerably. Seen here are two extremes: the one on the left loads the ring most highly at the gap, whereas the one on the right loads the area by the gap the least amount. The ideal situation is to have even loading all the way around the bore, though once gas pressures get behind the ring, especially the top ring, the differences in cylinder wall loading become academic.

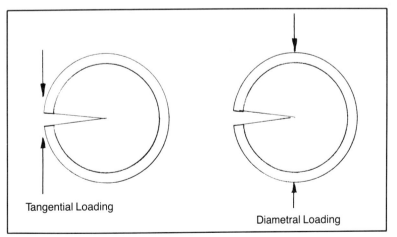

Tangential Loading

Diametral Loading

Ring tension is commonly quoted as the result of one of two ways of measuring it, either by tangentral loading or diametral loading. With the same ring, each method gives a different reading. With one ring measured both ways, the diametral loading gives the highest figure.

loss for the engine as well as a substantial increase in blowby.

If the groove is too narrow, the gas pressures will not build up sufficiently behind the ring and its sealing ability again will be limited. The correct balance between the two must be struck.

For production engines, groove clearances between 0.002-0.004in are

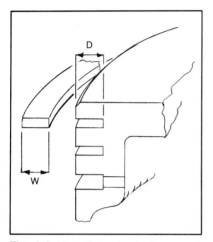

The definition of ring back clearance is the width of the ring (W) subtracted from the groove depth (D). The groove depth is measured from the OD of the piston at the ring belt; remember this is usually smaller by an appreciable amount than the skirt diameter.

used, but for a conventionally ringed race engine these clearances need to be kept on the low side of this tolerance at around 0.002in. Some top engine builders use as little as 0.001in but if they go much under this, problems can arise with rings sticking in the grooves as soon as any solid combustion products reach that area.

When vertical clearances are held to a minimum, back clearance behind the ring becomes more critical. If there is too much volume contained behind the ring, it takes too long for the pressure to build up to a suitable level to push the ring against the bore. Under these circumstances the volume contained in the groove behind the piston ring needs to be strictly limited. A groove back clearance of 0.020-0.040in is common on a production piston, but for a custom item it should be less.

The ring belt is much smaller than the piston skirt or the bore, so when the rings are in contact with the bore the static back clearance is increased. Why have all that clearance? Basically it is needed to accommodate piston rock. As the piston goes over TDC it can rock sufficiently for the top ring land to actually touch the bore. This means the ring can be pushed into the groove until it is virtually flush with the piston's ring belt OD—one more reason why a

minimum piston-to-bore clearance is desirable. On the custom race pistons, back clearance, as measured from the OD of the ring belt, typically is kept down to 0.004-0.005in.

Piston Gas Porting

The tighter the ring's vertical clearance in the groove, the more difficult it is for gas pressure to get behind the ring. This is where gas porting the piston is important. By drilling holes through the crown to join the groove, gas pressure can be communicated from the combustion chamber to directly behind the ring.

There is no doubt that gas porting a piston can work. It has proved an effective method of sealing up the engine; however, it does generate considerably higher friction at the ring face during peak combustion pressures. This can cause greater ring wear, but more importantly, accelerated bore wear right at the top of the bore.

The higher friction level may appear to be a disadvantage as far as power is concerned, but it occurs at a point where the rod and crank leverage is high. Therefore the drag, as seen by the crankshaft, is minimized due to the leverage it has over the piston. In other words, the crankshaft turns quite a few degrees for only a small amount of piston movement, hence the increased friction level has little effect on the crank.

The question of cylinder block wear arises here and is one which any racer with less than an unlimited budget must address. It is usually necessary to keep the block operating as long as possible and so it is wise to think twice about gas-ported pistons. Even if you're looking for maximum power, it is debatable exactly how much power increase a gas-ported piston delivers over a conventional ring. Understandably, gas ports can plug up with combustion deposits, making them applicable to drag-race or sprint engines only.

Ring Width

For a drag-race engine a 0.043in top ring seems to be the way to go for most applications, especially if the engine is to turn more than 8000rpm. For super-high-rpm applications, some engine builders even use a 0.030in wide ring. For an endurance engine or circle-track motor, top ring life needs to be consistent with the amount of usage and thus a 0.0625in ring seems

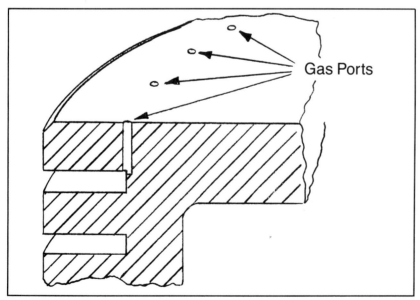

There is a possibility that rings, which fit snugly in the groove with minimal vertical clearance, may be sluggish in operation due to the difficulty of gas pressure getting behind the ring to push it out on the bore. Gas porting the pistons, as shown here, cures that problem. Increased bore wear, however, makes gas porting suitable for drag-race operation only.

to work every bit as good as the thinner ring, but lasts much longer.

If you are especially concerned about bore wear and plan to turn the engine at high rpm, a Dykes ring is a good route. Dykes top rings are preferred for test engines expected to run extended periods and for engines turning high rpm such as in drag racing. The Dykes ring is easy on cylinder bores because of its increased width, yet seals well because gas pressure backs it up instantly during combustion.

Ring width also determines the amount of combustion charge left unburned in the ring land and passed out of the cylinder on the exhaust stroke as wasted fuel and polluting emissions. By halving the ring land volume, significantly less charge is exhausted unburned. Sealed Power recognized the problem of unburned gases trapped in the ring land and developed its Headland ring.

The Headland ring is built to take combustion temperatures and pressures at the face of the piston crown. It's a heavy ring, but it is not so heavy that it cannot be used up to about 6000rpm. Most high-performance street motors run in sensible day-to-day use rarely need to turn more than 6000, so this ring represents a good compromise in terms of low ring wear, low bore wear and effective reduction of emissions as well as a potential gain in fuel efficiency. Moving the ring lands up from 0.300in below the piston deck to 0.150in can be worth as much as 10hp; this gain usually comes right along with a 1–2 percent improvement in fuel consumption.

These principles of ring position also apply for normally aspirated long-distance race engines. For drag-race engines, though, fuel consumption is hardly a consideration. Power is the sole criteria, so moving that ring as close to the piston's top, consistent with reliability, is what is needed. Just how close this can be will depend on the use planned.

If your plans include nitrous-oxide injection or supercharging, you must consider a different ring position. Under these conditions the piston crown experiences a far higher thermal load, causing the top ring land to get much hotter. To prevent the ring land from collapsing it is necessary to position the ring farther down the piston. A yardstick is to take 0.150in as the lower limit for long-distance or street engines, and add 0.010in for each pound of boost the engine is likely to use.

If you're using nitrous, try 0.010in for every 20hp extra that you expect from the engine. For instance, if you're going to inject the equivalent of an additional 200hp into the engine, a ring land position some 0.250in down from the top of the piston is a safe bet. The extra land width acts as a shield and extra conduction path to keep heat out of the top ring. If the top land is reduced in thickness, the top ring gap will almost certainly need increasing to compensate for additional thermal expansion.

Second Ring Position

The position of the second ring and the type of ring used is equally important. The second ring functions as a back-up pressure control ring to the combustion products, but it also can have an important secondary function: it acts as an oil scraper to assist the oil ring.

The position of the second ring from the top of the piston is not as critical as the position of the second ring in relation to the top ring. The land between the two rings has to support a certain load due to the inertia forces plus the gas pressure loads transmitted from the top ring. The second land, though, does not see temperatures as high as the top land. From that point of view it is unlikely to cause any distinct problems if it's made a reasonable width; something in the range of 0.150–0.200in is normal for a high-performance piston, although widths as much as a 0.250in are used on stock pistons.

If the application involves moderate temperatures or short operation times, the second ring land width can be cut to about 0.125in. If you're uncertain that the piston will cope with 0.125in land due to unknown thermal characteristics of the engine, then stick with nothing less than 0.150in.

Pistons with thermal barrier coatings can cut a substantial amount of heat getting into the ring land area and can usually run 0.125in width without problem.

Second Ring Width

There are a number of considerations for deciding on a ring width to use. First, consider the ring type as this can have an influence on the width of ring selected.

The second ring can also perform, if you so choose, the important function of assisting the oil scraper ring; this type of ring is a Napier scraper ring. The face of the Napier ring is slightly angled so that on the upward stroke it rides over the oil film, but on the way down it scrapes oil off the bore. The ring has a fine taper and initially, contact with the bore is made only on the edge of the ring, so the ring-to-cylinder-wall pressures are high.

The second ring has, by comparison with the top ring, little pressure backing it up to increase force against the bore. If it is a Napier scraper ring, then the tension of the ring is important as ring tension determines to a great extent how hard the ring presses on the bore, in addition to how much area is actually contacting the bore. For this reason the performance of a Napier scraper ring can decay quickly, because once a few thousandths inch has worn off the ring, it's no longer a taper ring—it's simply a plain ring, and it's scraping action is reduced.

As a second ring on a high-performance or long-distance race engine, the Napier has much to commend it. Made of the right material, such a ring can go through a 500 mile race and still only be contacting 0.020in or so of the width of the ring. Because only part of the ring OD is contacting the cylinder wall, one

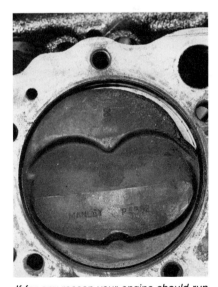

If for any reason your engine should run into detonation, here's what will happen if the ring land is too close to the piston crown. Note that the ring land has eroded away on the intake side rather than the normally expected exhaust side.

might assume that a 0.043in ring will suffice and save some space and weight; however, a 0.043in ring is going to need more built-in tension to be able to push it out on the bore; having lost some of its thickness, it's going to need more preload to generate the same radial pressure.

If you can work with a custom ring manufacturer, there is some small advantage to using the thinner second compression ring as opposed to a regular 0.0625in ring. But otherwise, 0.0625in Napier works well in the second groove.

Ring End Gap

To determine a safe minimum cold end gap, a Dwyer blowby gauge is needed. This is a standard piece of equipment that can be connected to an engine to measure blowby past the rings in cfm. For these tests no sump pan evacuation of any shape or form can be used, and the engine must be vented solely through the blowby gauge.

Now let's set a standard from which results can be judged. We'll employ a typical engine with a set of regular forged pistons, clearanced in the 0.005in range, and a 0.0625, 0.0625, 0.1875in ring set. The compression rings would normally be gapped at 0.018in for the top ring, and 0.014in for the second. For a small-block Chevy having had the bores deck-plate honed, blowby around 3-4cfm would be normal at operating temperature. But the end gap of both top and second rings can be reduced—by how much largely depends on the ring position in relation to the piston deck.

The higher up the piston the ring is, the hotter it will get, therefore the more cold end gap required. Assuming that leakage does not occur from other sources, such as a poorly fitting ring, bad surface finish on the bottom groove face or too much groove back clearance slowing the ring's response to pressure backing, then you can cut the ring end gap until the leakdown rates are around 1.5cfm. To get much under 1.5cfm leakage by reducing the end gap seems to be dicey, at least when we're considering a typical forged piston.

A close-fitting hypereutectic piston, which controls the ring position relative to the bore much better by virtue of a close piston-to-cylinder-wall fit, can produce a significantly better seal. Tests with hypereutectic pistons

at 0.0022in cold clearance and careful control of end gaps produced blowby figures down to 0cfm. Such low levels are not practical with clearances typical of a forged piston.

Zero-Gap Rings

Tests also indicate that near-zero blowby figures can be achieved using the sort of clearances expected for forged pistons when a zero-gap second ring is used. The particular rings tested were the Childs & Albert stepped rings. These are high-quality rings and performed well even when transferred from engine to engine to see how they lasted over time. The rings went through three engines and covered the equivalent of some 1,500 racing miles and still performed well.

Sealing in the cylinder is important, but the real issue is how much does it affect power output? There is always a trade-off between frictional losses and blowby. Performing back-to-back ring tests can be time consuming, and often the results have to be looked at statistically to determine a trend. A set of rings that are worn and leak but are low on friction can deliver as much power as a set that seal but have additional friction. The price for cutting blowby can often be increased friction and the end result may be no more power than before. How much effort should be applied to cutting blowby, and at what point does it become a futile exercise?

My current policy is that reducing blowby is the number one priority, so I strongly advocate the use of torque plates for honing. If getting the last drop of sealing performance didn't matter then it becomes necessary to question the validity of torque plates, but I feel the value of a torque plate is proven. It always pays to remember that bore friction accounts for the greatest single loss of power in an engine.

Most of the engines used in my shop are long-term dyno test engines. For such engines to be of any use as yardsticks, they need to repeatedly produce the same base-line power curve; the power curve needs to be the same whether the engine is on its fifth test run or 500th. In most cases re-base lining an engine is time consuming. Although on occasions repeated re-base lining is necessary, it's best to minimize base line testing.

To measure blowby it is necessary to vent all blowby through the gauge. This means disconnecting any

crankcase evacuation systems, which I use on 90 percent of my test engines. When I run crankcase evacuation it precludes any use of the blowby meter, so instead the crankcase vacuum is monitored. If blowby increases it obviously becomes more difficult to maintain a crankcase vacuum.

On a well-built engine, crankcase vacuum stays relatively constant for a long period of time, and power follows suit. However, when crankcase vacuum starts to drop, shortly thereafter power starts to follow suit. This is a strong indication that there is a reasonable correlation between blowby and horsepower.

Re-ringing an engine that has blowby may not produce additional power, because the new rings are also changing factors other than just sealing capability. With new rings, friction is almost certainly going to change—most likely in an upward direction.

During Billy Howell's tenure at Chevrolet, where his responsibilities were performance oriented, he had numerous experiences relating to blowby reduction versus horsepower. He has seen cases where blowby in excess of 10cfm was cut to 2½-3cfm with little or no measurable increase in power. He makes a valid point that the negative effects of blowby are seen more in perspective by looking at the cfm of blowby in relation to the cfm the engine inhales. For instance, if a motor consumes 400cfm at 5000rpm, and blowby is 8cfm of hot gases, then the 8cfm would at the same temperature as the intake, be about 4-5cfm. This represents a 1 percent loss of charge mass. Would it be fair to say that fixing this should at best result in a 1 percent increase in power? This is difficult to test for, but it seems to be a reasonable assumption.

There are, though, some side issues which have an effect on the final outcome. If an engine has excessive blowby, windage losses are increased because the oil stays in suspension easier. Also, oil aeration is increased. The bottom line to test for the effect of blowby reduction is to introduce a measurable artificial leak from the cylinder, rather than use what goes past the rings. Testing this way can isolate blowby from all other factors and allow a true relationship between blowby and power to be produced. A simpler test, though not necessarily one which gives a truly accurate picture, is to measure horsepower versus blowby as a given set of rings beds in seals up to its maximum, then wears out. This reduces the number of variables but certainly doesn't isolate the effect of blowby.

Although back-to-back tests have not been performed on zero-gap rings, building of similar engines has indicated that zero-gap second rings seem to make additional horsepower. Some builders claim increases of 12–14hp; my shop dynamometer indicates an increase in the 5–7hp range. Even so, such claims must be made with a certain degree of reservation.

The conventional rings against which the zero-gap rings were tested were gapped at a dimension known to be safe. In other words, neither temperature nor dynamic conditions would cause the end of the rings to butt together. If ring gaps are too tight, the ring will expand, the ends will butt together and the ring will bind in the bore. This will cause either a seizure or at least a tightening of the engine and a subsequent power loss.

If you are repeatedly building an engine of essentially the same specification, the key is to progressively close the end gaps until signs indicate the rings are butting. At this point, increase the gap slightly to a safe dimension. If the minimum ring end gap is found, then any gains due to the use of zero-gap rings will be minimized.

Trying to find that elusive minimum end-gap dimension can be an expensive lesson because substantial ring binding can occur just by reducing the end gap an odd thousandth of an inch or so. Assume that the rings you have are already gapped so that under working conditions, the ring expands, reducing the end gap to just shy of zero. Not knowing the gap is near zero, the next engine built has the rings gapped a little tighter so that as working temperature is reached, they butt. This will cause ring friction on the bore to increase dramatically, leading to higher ring temperatures. This additional heat causes the ring to tighten up even more. At this point, things can spiral downward by damaging or destroying the block plus the pistons. To avoid these events, it often proves cheaper to go for a zero-gap ring, even though a set of eight may be twice the price of a regular ring set.

The zero-gap or gapless style ring would seem to be an obvious means of completely sealing up the cylinder. However, one theory is that if the second ring seals effectively and prevents any blowby whatsoever, then leakage past the top ring will cause gas pressure to build between the top and second ring. This means that after the combustion pressure is released, the pressure underneath the top ring will cause it to unseat from the bottom of the ring groove and leak—obviously something you want to avoid.

Part of the ring's ability to seal hinges on the fact that an ultra-thin layer of oil is trapped between the face of the groove and ring. If the ring repeatedly unseats itself, the passage of gases will scour the oil film from the surfaces, reducing the seal and increasing wear. The worst situation is for the seal between the ring and the lower groove face to fail immediately prior to the power stroke.

Buffer Grooves

Because of the leakage problem and the subsequent pressure build-up between rings, some builders put a buffer groove between the top and second ring. The thinking here is that any leakage past the first ring will disburse into a relatively large vol-ume, and therefore, the amount of pressure build-up will be reduced. Since the pressure build-up is reduced, there is less likelihood of the top ring lifting from the bottom of the groove. That seems pretty reasonable, however, things are not quite that simple.

It has been found that with an effective top ring seal, the pressure will build up in the buffer groove because the blowby back out from it is slower than the blowby into it—unless it lifts the top ring from the bottom of the groove. The power stroke is quicker so the amount of time available to release trapped pressure is longer, but having the groove there is not a guarantee that the ring will not lift from the bottom face of the groove. On the other hand, if the ring does *momentarily* lift, but reseats itself prior to the power stroke, then there's little to worry about.

There are two ways to look at a possible solution here. The first is to increase the volume between the rings with the buffer groove, and hope that the pressure never gets high enough to unseat the ring. Conversely, the volume can be made minimal so that any pressure that does build up will quickly leak away. For what it's worth, tests have shown

Two things to note on this piston. First, the buffer groove between the top and second compression rings. Second, note that the oil ring groove passes through the pin bore. After the piston has been assembled on to the rod, a support rail is loaded into the groove to provide a surface for the oil control ring to ride on.

zero blowby both with and without a groove between the top and second ring. The bottom line here is that for the second ring to be effective it is necessary for the top ring to operate as effectively as possible.

We are now ready to install a ring package that has virtually zero blowby, so everything should be fine and dandy. Well that may be so, but there's more to the situation. Bill "Grumpy" Jenkins, in his book *The Chevrolet Racing Engine* (S-A Design) states that an engine needs to have some blowby as a major oil control factor. If we have to choose between oil getting up to the combustion chamber or blowby coming down past the rings, the blowby, so long as it's not excessive, is by far the lesser of two evils.

To make his point, Jenkins modifies his pistons so they have an oil return groove originating from above the oil ring. A drilled hole from the pin bore connects to this groove. Oil collecting in the groove migrates to the hole and pin bore by way of blowby escaping past the top and second ring.

If there is a distinct quantity of oil above the oil control ring, this indicates that the oil control ring itself is not providing 100 percent control of lubricant on the cylinder walls. There are plenty of indications that this is

occurring because a Napier scraper, as opposed to a plain ring in the second groove, noticeably aids oil control. Further proof would be to install a Napier scraper ring upside down and see how much the engine becomes an oil pumper. Remember, the second ring can only pass oil up the bore that was left by the oil control ring in the first place. From this we can safely conclude that oil control is accomplished largely by the combination of the second ring and the oil control ring.

Since the Jenkins oil control groove and drilling picks up oil above the oil control ring, we also can conclude that it is mostly picking up oil left by the oil control ring, but prior to the interception of the oil by the second ring. If the second ring is acting as a scraper pulling oil down, much of the oil that would have to have been dealt with by the oil control ring should now pass out through the groove and down through the pin, lubricating the pin in the process. There are pros and cons to this method, but that's a subject to deal with later.

Apart from the routing of the oil hole, this technique has merit for aiding oil control. I think that it may be better to pass oil from the bore walls through the piston in some route other than that of a restrictive pin bore; a more mundane pin oiling

system may prove adequate. This would allow oil removed from the bores to be more freely routed back to the pan via simple oil return holes.

Oil Control Ring

Most of the heat of combustion has been dissipated prior to reaching the oil control ring. Frequently an oil control ring sees about 250deg. Fahrenheit, meaning that the aluminum is stronger and so the amount of space we need to devote to the third ring land is much less than the other two. For a piston of good-quality material, as little as 0.070–0.100in can be used. But for the most part, a wider ring land is used, typically 0.125in.

Most rings used for oil control are multi-piece units comprised of two rails and the support structure. The two rails are located at the top and bottom of the groove by its supporting structure, and the assembly fits snugly into the groove. When considering compression rings, the point of a thinner ring is to reduce the area in contact with the bore so that it seats in more quickly and hopefully better, and to reduce inertia in direct proportion to the reduction of ring width. With the multi-piece ring, dropping the ring from 0.1875in to 0.125in does not reduce the ring weight in the same proportion. In fact, a 0.1875in ring weighs only marginally more than a 0.120in (3mm) or 0.125in ring. The principal argument for using the thinner ring is to allow the ring pack to be smaller, thus accommodating a longer rod.

One of the most effective oil control rings is the Sealed Power SS-50 ring, available as a 0.125in or 0.1875in ring. Another quality ring is the Hastings, available in comparable widths but popularly being used in a 0.120in width on Wiseco pistons.

All these piston rings can be made to work effectively, but there is a key to their operation. First, ring tension, as applied to the bore, has a noticeable effect on how much oil is consumed. However, the higher the ring tension the more friction there is, and lower tension rings definitely assist in the production of power. Tests by top Winston Cup engine builders indicate that there is 5–10hp in the oil control rings.

If the ring tension is reduced, oil control can get significantly worse unless precautions are taken. The key to oil control in a small-block Chevy

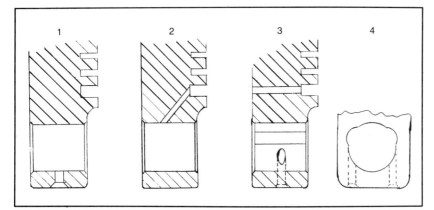

The wrist pin can be oiled in various ways, some that are better than others for high-performance applications. Pistons lubricated by simple holes drilled in the base of the boss (1) are the most prone to cracking at the hole, simply because it's drilled in the thinnest place of the boss. One of the most common techniques for high-performance pistons (2) is to drill a communicating hole between the pin and the oil ring groove, so oil collected by the ring is passed down into *the pin. Piston accelerations can cause a considerable pressure in this hole to drive the oil between the pin and piston. A third alternative (3 and 4) uses a combination of systems, and one or other will suffice. If there's inadequate pin oiling, two holes can be drilled in the base of the piston boss. Some pistons, however, use oiling troughs, these are reliefs cut in the pin boss at about 10 o'clock and 2 o'clock.*

engine is to make sure there is a positive pressure differential existing at all times between the cylinder and the crankcase; this means that the crankcase must never be pressurized.

In fact, it is better to pull the crankcase pressure below atmospheric pressure so that there is a tendency for oil to migrate down the cylinder walls rather than up. In essence, low-tension rings should only be used with engines that have some form of crankcase evacuation. If we're dealing with a typical street motor, then regular-tension oil control rings should be used; otherwise, needlessly high oil consumption may result.

There are several advantages in a 0.120in ring or any narrow oil control ring over the conventional 0.1875in size. Some engine builders claim that the smaller oil return holes of a 0.120in oil ring cannot discharge the oil fast enough back to the crankcase. In most cases this should not be a problem. If adequate oil holes are put into the back of a 0.120in or 0.125in ring groove—and they need not be excessive—then oil control can be maintained. This assumes that piston clearances are suitably controlled to hold the ring in position in relation to the bore, and that sufficient crankcase evacuation is achieved.

When dealing with a race engine, the amount of oil at the top end of the engine *should be restricted* as an all-out race engine is almost certainly going to employ roller rockers. Of course, lubrication of the top end should not be restricted to such an extent that there isn't enough oil to cool the valve springs. Otherwise we may have a powerful engine, but it's no good in a 500 mile race if you break a valve spring at the 300 mile mark due to heat and fatigue.

The question of oil control in a high-performance engine is a delicate one. Having plenty of oil on the underside of the piston is good for cooling but not good for oil control as excessive amounts will collect at the cylinder bores. This will lead to greater losses from ring drag as the rings have to scrape more oil off the bores. Ring drag in a typical high-performance engine accounts for about 12 percent of the mechanical losses of the engine, whereas the remainder of the piston assembly accounts for about 20 percent of the losses.

This is an area where attention to rings can pay dividends, but equally important is that attention to losses caused by the piston skirt form and other factors can also pay dividends. It's all a question of how easily the rings can scrape the oil off the bore on their passage down the bore, and how easily the crank cuts through the oil. Proof of this is that the thinner the oil is, the more horsepower an engine will make.

Regular-pressure oil control rings tend to have a radial pressure of between 20–22lb; low-tension rings tend to be around 15–18lb. A high-tension ring is a pretty safe bet for controlling oil under any circumstances and is the type used for street motors from production rebuilds to race engines. The low-tension ring can be used on street motors but the circumstances have to be right; otherwise, higher oil consumption can result, but the low-tension ring is the way to go in a motor that has an evacuated sump.

Wrist Pin Location

The wrist pin position in relation to the oil ring groove has been a controversial subject over the years. For a small-block Chevy engine builder it is a subject worthy of consideration, as larger displacement engines need to accommodate as long a rod as possible to achieve a favorable rod-to-stroke ratio. Since the block is really too short for anything much over a 3.250in stroke, you need to put the wrist pin as high in the piston as possible.

Normal practice is to put the pin bore up close to the oil control ring. But demands for even longer connecting rods force engine builders and piston manufacturers alike to look for ways and means of allowing the pin bore to encroach into the oil groove, yet still have a functional oil control ring.

One way to achieve this was to put plugs into the ends of the pin holes and machine these in conjunction with the oil control groove, so that part of the groove was in the plug. Then the pin actually could be moved up into the area of the oil control ring. Though feasible, this practice has gone by the wayside.

The current technique for putting the pin up into the ring pack is to use a support rail on the bottom of the oil control ring. The piston is assembled onto the rod, and the pin located in the bore and whatever retention method used is fitted to hold the wrist pin in place. This rail that grips the bottom of the ring groove is thick enough to act as the lower face of the groove and bridge the gap where the pin passes through. The oil control ring simply fits between this rail and the top of the groove. The technique seems to work well and allows the pin to be moved up about 0.125–0.1875in farther than would otherwise be the case.

Wrist Pin Bore

We've talked about wrist pin position in the piston in some depth already; however, this has all been in relation to how long a rod we can get into the engine. There are other factors that need to be considered. Getting a super-long rod into an engine only becomes of prime importance if the stroke of the motor will not accommodate the rod length required to get a good rod-to-stroke ratio. Obviously, a small-displacement engine with a short stroke is going to be able to get a long enough rod into the engine without the need to push the pin as near to the crown as possible.

When the pin location is not dictated by the need for a long rod, the pin is located about the middle of the skirt area so the piston has the least tendency to rock; however, this can only be done with pistons for engines of some 3.00in stroke or less. Assuming that a pin in the middle of the

Oil return holes do not necessarily have to be at the bottom of the oil groove. Indeed, there are a number of benefits from putting oil holes below the bottom oil scraper ring, as this allows the bottom scraper ring to function more efficiently.

Expansion and Cold Piston Clearance

Have you ever wondered how much pistons expand in operation? How much of that cold clearance is left when the engine is at operating temperature? It's not difficult to calculate a fairly accurate figure.

For most pistons made of a low-silicon alloy, the expansion coefficient is 0.000013in per inch per degree Fahrenheit. This figure may vary depending upon the alloy involved, but it is a good ballpark figure.

To get an idea of how much hot clearance the piston will have, it is necessary to know the cold size of the piston. Since aluminum expands quickly with heat, it is important to establish its cold size at a fixed temperature. In engineering it is common to perform measurements at a standard temperature of 68deg. Fahrenheit. Therefore, to establish the cold

piston-to-cylinder-bore clearance, the pistons should be left overnight to heat soak at 68-70deg. before being measured. This is necessary because the change in temperature from a cold to a hot workshop will measurably affect piston size.

The operating temperatures of the skirt increase by 180deg. over the nominal 68deg. measuring point. Knowing this and the expansion coefficient for the alloy, skirt growth can be calculated. On a typical 4in piston this would be equal to 0.000013in times 4in times the temperature rise of 180deg., which equals 0.0094in.

In addition to piston expansion, we must also allow for bore expansion with rising temperatures. Cast iron, though, doesn't expand as quickly as aluminum, and the average temperature of the cylinder walls is going to be lower than that of the pistons. Typically the cylinder wall is going to run around

200deg. and cast iron with an expansion rate of around 0.000007in per inch per degree Fahrenheit will expand 0.0035in.

Thus our example indicates that whatever the cold clearance is under hot running conditions it will, for a low-silicon alloy piston, decrease by approximately 0.006in.

Any design feature that makes the skirt of the piston run colder or hotter will, of course, ultimately affect the running clearance. For instance, a thermal barrier on the piston crown can make the skirt of the piston run 25-50deg. cooler. Under these circumstances the cold clearance can be reduced. Pistons with a slot in or below the oil groove cut the heat conducted to the skirt, so they can be run with a tighter cold clearance.

All these factors have to be taken into account if the optimum hot cylinder wall clearance is to be achieved.

Here's a piston from a race-winning Winston Cup motor. At various points the crown has been hardness checked. Knowing the hardness of the piston to start with and the temper curve of alloy, hardness checking can reveal the average temperatures that each particular area runs at. Notice the softest, and thus the hottest, part of the piston is directly under the exhaust valve at the peak of the crown. Also note that the point at which detonation takes place, which is at about 10 o'clock on the intake side, the piston is relatively hard. This piston has not been detonated, so until detonation actually occurs, this indicates that the intake side of the piston actually runs cooler than the exhaust side. This may be an indication that the location of the spark plug in the cylinder head needs to be moved nearer to the 10 o'clock point on the piston to produce a more evenly radiating burn.

Surface coatings on the crown of a piston can significantly reduce the temperature of the parent metal immediately beneath. This results in a cooler running, physically stronger piston.

skirt is ideal, how much of a trade-off is there when the pin is pushed up into the ring pack for rod length in longer stroke engines? Current experience indicates that even though the ideal pin position may be in the middle of the skirt, the need for a longer rod in a long-stroke small-block Chevy is greater, so moving the pin up as high as possible in order to get in a long rod pays off in terms of both engine output and increased bore life.

Piston weight is an aspect a racer always needs to consider. In an effort to cut weight, wrist pins, being made of steel and one of the heaviest parts of the piston, are pared down to a minimum. Of course, cutting them to a minimum weight can also reduce reliability and they almost always fail in bending. To keep bending to a minimum it's necessary to keep the piston support piers as close as possible to the rod. The pin is subjected to less bending load at the points where it is held in the piston or rod. Thus, it's necessary to get the sides of the piston as close to the sides of the rod as possible, allowing enough side float to make up for misalignment.

Still, the amount of side clearance between piston and rod is dictated by the rod builder and the forging; an end user has little control over it. Though side clearance between pin towers is largely beyond control of

the engine builder, the pin fit in the bearing surface is not, and under some special circumstances this can be a potential problem area.

Four factors affect the fit necessary for survival of the bearing surface between the piston and pin. First is the power loading the piston is expected to take. The second is the rate at which the piston warms up, as this affects whether or not sufficient heat has reached the pin bore to expand it to the required working clearance. Third is the method of pin oiling. Finally, there is the effect of the piston material's expansion coefficient; high-expansion alloys obviously require less cold pin clearance than low-expansion alloys. Not surprisingly, all these factors are closely interrelated.

Most production pistons are made with close pin clearances, which eliminates small-end rattle. Typically, a production-cast piston will have around 0.0002in pin clearance; most low-expansion forged pistons such as the TRW and Sealed Power are of the same order. The cold clearance used for most race pistons made of low-silicon alloys, which usually have higher expansion ratios, start at about 0.0002in but because of higher heat input and expansion rates, their running clearances tend to go to 0.0015–0.002in.

For most street performance engines, pin clearance is something

Here's a variety of pin sizes with their weights inked on them. As you can see, the difference in weight between the lightest one and a stock pin weight, which is around 150-155g, is substantial.

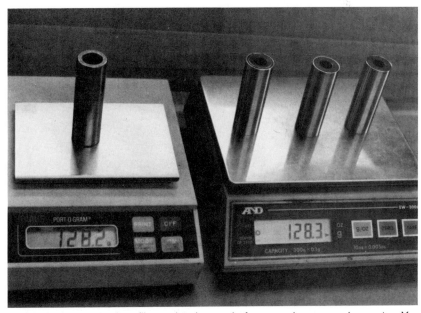

At right is the new carbon fiber wrist pin from PRD, which is much lighter than the lightest steel pin. On the left, is a typical pin for an endurance-racing motor. You can see that the equivalent carbon fiber pin weighs three times less.

you'll never need to pay attention to. But if heavy loads of nitrous are used in a modified motor, then things change. The TRW and Sealed Power forged pistons are capable of handling high power outputs, but certain precautions need to be taken if these pistons are to be used at much above 500hp. As of 1990, not all the forged pistons from these two manufacturers have positive oiling to the pin; positive pin oiling refers to oil forced to the pin via the oil control ring groove. Some pistons simply have a plain pin bore.

Pistons with a plain pin bore are going to need better oiling, plus more clearance, if the pin bore is to survive. Nitrous motors in the 600hp range will need as much as 0.0008in of cold pin-to-piston clearance to survive. This will be especially true if the engine is used for drag racing.

A point worth noting is that the engine should, as far as possible, be warmed to bring the pistons as near operating temperature as possible. Of course, they never will be at operating temperature no matter how much it is warmed prior to making a pass. So why the need to heat soak the pistons? It's because nitrous can dump so much heat into the piston crown so quickly that it expands

appreciably before much of that heat has flowed to the rest of the piston.

During this time two things happen: First, the pin bosses get out of alignment due to the crown spreading; and second, most of the heat hasn't had time to get down to the pin bosses and expand the pin bore to a suitable running clearance. Consequently, the increased loads, plus a tight pin, increase the possibility of a seizure—hence the need for the extra cold clearance. This extra clearance seems only to be necessary for high-silicon, low-expansion pistons. If pistons are purchased from a specialty manufacturer and the use of nitrous is intended, then it's wise to consult them about pin clearance for their particular piston.

Pin Oiling

Pin oiling can be accomplished in a number of ways. A typical stock-type piston has a plain bore with the hole drilled in the base of the bore to pick up oil splashed from the crankshaft. Oil simply migrates along the pin-to-bore interface from this point. For most street applications this is adequate, but for a high-output racer it has limitations. First, adequate oiling may not take place; and second, the drilled hole is normally at the thinnest and most highly

loaded part of the lower side of the boss. This forms a stress raiser and with a piston designed for minimum weight, the boss could crack at this point.

The second piston type has oil returned from the oil scraper ring groove passed down to the pin bore. On the Jenkins piston mentioned previously, the oil drilling connects a groove above the oil control ring to the pin bore. The only problem with both these types of oiling is that it passes hot and possibly contaminated oil down to the pin boss. The oil scraped from the cylinder walls can contain combustion deposits and metal particles.

In practice such pin oiling systems are effective, but other techniques can be used. For instance, both TRW and Cosworth are using oiling troughs or reliefs. This involves relieving the pin bore with a shallow depression located at a critical position in relation to pin flex, a subject we'll cover later. Oil splashed up the crankshaft finds its way along this groove and migrates around the pin. The benefits of this system are that it is easy, effective and does not cause any significant stress raisers.

It's not uncommon to buy a set of off-the-shelf forged pistons that need some help in pin oiling. The simplest technique is to just drill the pin boss on the pin centerline at the bottom of each pin piece. But this can cause a fatigue problem.

Since most pistons have square pin bosses on the underside it is better to drill two holes that just break into the side of the pin hole. These holes

The wrist pin bore on this piston shows every sign of being too tight for the nitrous application for which it was used. Heavier piston loads need to have wider pin clearances to accommodate pin bending and to allow more lubricant to pass through between the bearing surfaces.

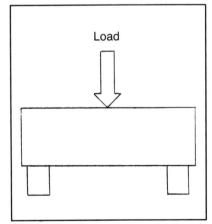

If simple beam loads, as shown here, were all that were involved, designing a wrist pin would be easy.

drilled in the thickest part of the pin boss are less likely to cause fatigue failures and provide superior oiling.

Press-Fit Pin Retention

In stock form a small-block Chevy wrist pin is press-fit in the rod. If the intention is to build a street motor that will stay together for any length of time, the press-fit method is as good as any. There is no horsepower advantage to having a floating pin in the rod; however, there can be some disadvantages. If the stock rod is retained, then the press fit that already exists is convenient. If the stock rod is retained but for some reason a floating pin is desired, then this will require changes in the rod.

The most common technique is to install a bronze bushing to reduce wear to a minimum. Some engine builders make the connecting rods fully floating by simply honing the rods out and running with the steel-on-steel bearing. It is possible to get away with this under certain conditions, but for a street motor it's not the best way to go.

If a bushing is used, then material must be machined from the small end of the rod and this can weaken the rod at that point. Whether or not this becomes a major problem depends on whether the rods have enough material around the pin boss for a safety margin when the bushings are installed.

Floating Pin Retention

It would appear that since no real side loading should occur on the wrist pin, floating pin retention would be simple and nothing particularly sturdy should be needed. However, this is not the case; there can be strong pin side loads generated from many sources.

First, there can be momentary and uneven pressure build-up across the width of the piston with the pressure in the cylinder building up and spreading across the combustion chamber with the speed of sound. This may appear to happen quickly, but it still takes a finite time and for a moment the pin can be loaded more on one side than the other. Also, the rod is slightly offset from the crankshaft journal. This can lead to side loading. Any errors in the piston, bending in the rod, or misalignment of the big-end journal with its small-end counterpart can also generate a degree of side loading. Finally, if

the crank journal itself is not square to the mains, which can be common due to various less-than-acceptable crank-grinding techniques, additional side loading can occur.

Full-skirt pistons tend to have less problems with pin side loading, as opposed to slipper pistons, but side loading does occur nonetheless. What happens is that the friction level between the wrist pin in the small end is enough to grip the pin and hammer it against whatever locking medium is being used. If a single circlip groove is used it is possible for the pin to hammer the circlip out of the groove.

Precautions against this happening can be taken by keeping the clearance between the end of the pin and the circlips to a minimum. For the most part, on engines I've built the pin length is a snug fit between the clips so there is little or no clearance when cold. This means as the piston warms up, it develops a few thousandths inch side clearance and keeps the hammering action on any circlip retainer to a minimum.

Because of the possibility of hammering out a single circlip, a popular method to eliminate this action is to use two circlips in each groove. This technique, known as double locks, seems to work quite satisfactorily; however, with both the single and double circlip it is necessary to make sure the circlip has seated properly in the bottom of the groove. If it is not in the bottom it won't be long before it hammers out and the pin, or the circlip itself, destroys the bore. Overall, though, double locks are simple to install, and reliable if inspected thoroughly at assembly time.

An alternative to the circlip is the Spiraloc. This is a spiral clip wound into the groove; it is almost impossible for it to dislodge.

The main problem with Spiralocs, though, is the difficulty of assembling and removing them. Experienced engine builders manage to make their installation and removal into something of an art. For engine builders without this daily practice, however, Spiralocs can be a challenge.

A third type of pin lock, popular in Europe and subsequently introduced on small-block Chevy pistons by Cosworth, is the round wire lock. This is probably the easiest system to install and in terms of race reliability delivers good results. Yet it does work differently from the regular method of pin retention.

For a wire lock, the end of the pin must have a chamfer. The retention action is actually reinforced as the pin impacts into the lock. Since the taper drives the lock harder into the groove, this has a tendency to ensure the wire never leaves its position.

As for assembly, the wire lock is probably the fastest type to assemble and disassemble. A groove cut into the pin bore allows a scriber point to be inserted under the wire lock and simply removed.

Wrist Pins

All stock and stock replacement pistons come equipped with wrist pins strong enough for almost any application. Assuming this is true there are only two reasons why you would make a change: either the type of piston used may need a different length pin, or the stock pin is too heavy. With a length of about 3.00in to suit stock-style piston, stock pins, at about 150gr, are unnecessarily heavy for most applications.

Unless you have set aside a budget specifically for lightweight parts, there is absolutely no need to make a change if rpm is kept below 6500. The stock pin's weight will not overload even stock rods up to 6500rpm. Above 6500rpm, the piston assembly's reciprocating weight becomes an increasingly important factor especially if a stock rod is to be used. This means the only reason for changing the stock pin is, in essence, to decrease the piston assembly's weight.

The irony is that the higher the horsepower and rpm potential of the engine, the more pin strength required. The only factor countering this is that most stock pins are substantially overdesigned; the manufacturers of stock-type pistons are conservative as to the amount of pressures and deflections they allow the pin to see.

Most pistons are designed so the piston-to-pin surface pressures do not exceed 8,000psi. A typical stock pin in a stock-style piston has about 2sq-in of piston-to-pin interface supporting peak combustion loads. Since it is not intended for extended use, a racing piston and pin can tolerate loads up to 20,000psi. Since peak cylinder pressures may only go up about 50 percent, in many instances the load-supporting area can be reduced, often by using a shorter pin. This may seem like a backward step; however, a shorter pin has two advantages: it's less prone to bending, and it's lighter in direct proportion to the amount it is shortened.

To see how pins react to loads, let's consider the kinds of deflection that occur in a pin. Initially there are bending loads considered to deflect the pin since it is centrally loaded. Pin deflection in bending causes two things to happen. The bearing surface loads at each end of the pin bore increase greatly because the pin is not riding parallel to the bearing surface. Additionally, bending of the pin causes the piston bosses to deflect and in many instances this induces early pin boss cracking. Pin bosses are often reinforced to prevent cracking, when in fact excessive pin deflection is the culprit.

For most stock designs a pin deflection of 0.003in is considered the maximum if the piston is to survive for high mileage. In practice, few stock Chevy pins bend this far for most of their life; it is only at full throttle and the engine's peak rpm that stock pins begin to approach this limit.

Because stock pins are tough items, producing something significantly stiffer is difficult. For short-term operation such as racing, higher pin deflections can be tolerated in an effort to save weight, but shortening a pin automatically increases its stiffness.

In practice, pin bearing lengths down to 0.625in per side can be utilized for endurance and drag-race engines—as long as the pins are adequately stiff. This means we should consider pins of 2.50in in length as near minimum. Street motors come under the category of endurance engines.

In considering what wall thickness is needed, the bottom line is the amount of pin deflection that can be tolerated under working conditions. If engine rpm and horsepower potential is limited by the mandatory two-barrel carburetion, the pin can have a relatively light cross section. To a certain degree the same would also apply to a drag-race motor, as it is only expected to make a few dozen passes before pistons and pins are changed.

If the engine is a more than 600hp unit for 500 mile races, then you'll need to consider component fatigue. Continual bending of the pin back and forth may not only cause increased wear at the pin bosses, but could also lead to a catastrophic failure. Part of the increased wear is due to the deflection of the pin in its beam-load status, and part is due to another factor known as the Rothman ovalizing deflection.

Under load, the end of the pin ovalizes, taking on a semi-triangular form if the deflection is exaggerated sufficiently. In a typical close-fitting pin-to-bore situation, as found in a street or high-performance street piston, the maximum amount of ovalizing that can be tolerated for long piston life is 0.0008–0.001in. The thinner the wall section of the pin, the more likely the end of the pin is to ovalize.

A compromising situation arises here. Initially at least, it appears a lightweight pin is best designed with the main mass of material at the center and with the section thickness decreasing toward the ends. This makes the pin strongest, for a given weight, in the beam-loading mode.

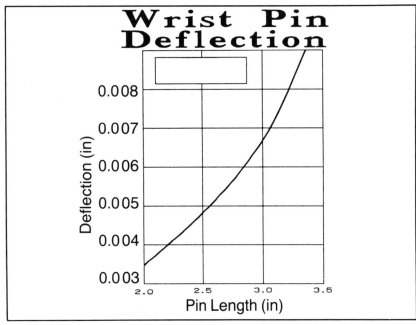

Here's how wrist pin deflection goes up with pin length. As you can see, the 2.5in pin is quite a bit stiffer, and suffers less deflection than the 3.0in pin.

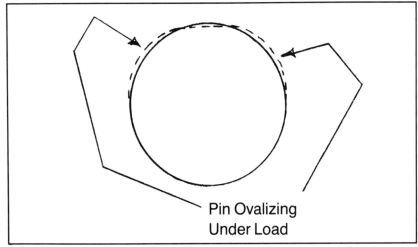

The dotted line here indicates the deflection at the end of the pin, known as Rothman ovalizing. Taper-wall pins are far more susceptible to this than parallel pins.

60

However, such a pin design produces the maximum amount of ovalizing at the ends. In this context, tapered pins are not necessarily the best way to go unless precautions are taken.

I mentioned that side reliefs for pin oiling need to be placed in strategic positions to take into account pin deflection. By relieving the pin bores in the appropriate spot, provision is made for the pin to ovalize into the relief grooves. This reduces the tendency to generate excessive surface bearing pressures between the pin and pin bore, counteracting the negative effects of pin end ovalizing.

At this point there is a dilemma in pin design: a taper pin is the best per weight for beam deflection; a parallel pin is the best in avoiding end ovalizing. The Roper pin design created by Cosworth's Jeff Roper minimizes both beam and ovalizing deflection.

The ends of the Roper pin are reinforced by making the ID considerably smaller. Ideally, the form of the OD of the wrist pin needs to be something other than an accurate parallel diameter. By having a shallow taper, the high end loads that occur when the pin is deflected can be reduced.

Pin diameter is also an important factor. The 0.927in diameter of the stock pin has become something of a tradition for the small-block Chevy. Few efforts have been made to change to an alternative diameter and you may ask why it would be necessary to do so. The fact is that as the pin diameter increases, the pin gets stiffer. Also, making the pin larger means more surface area is available, so the pin can be shortened. A large-diameter pin can thus be stiffer and lighter than a small-diameter pin, which must be longer to get the bearing area. All this is fine in a short-stroke engine where there is room beneath the piston crown. But with strokes over 3.25in, accommodating anything but the stock pin diameter is a problem.

To get an appropriate or maximum rod length into the engine, the pin has moved up the piston as far as possible. Under these circumstances a bigger pin is nothing but an embarrassment. Nonetheless, the larger-diameter pin is worth considering.

If the intention is to build a high-rpm drag-race motor around 300ci with as light an assembly as possible, then a 1–1.0625in pin could shave 10-15 grams off the pin weight with no sacrifice in strength. Since the pins are steel and comprise a major proportion of overall piston assembly weight, a reduction here makes a difference.

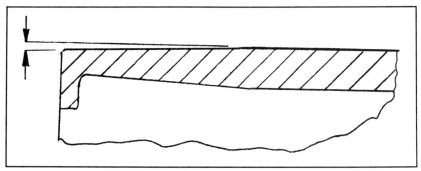

As a wrist pin bends, it causes higher stresses at the end of the pin. By having a shallow taper on the pin, more uniform bearing surface is generated at high loads.

The three types of pins you see here have various attributes. Pin 1 is relatively good in beam strength, and gives minimal ovalizing at the ends for a simple design. Pin 2 is much better in beam strength, but sacrifices strength at the end, and suffers more Rothman ovalizing. Pin 3 is a design intended to give the best of both worlds. Such a design is high in beam strength, and the reinforcing at the ends minimizes ovalizing.

Wrist Pin Selection

Pin Length (in)	Wall Thickness (in)	Suggested Application
2.50	0.090	Light piston usage. Good for circle-track motors where compression and breathing are limited; 2-BBL 9:1. Bracket-race motors limited to about 500-525hp.
2.50	0.120	Light piston usage. Circle-track motors to about 550hp short-race distance. Rpm to 7500. Drag race to 600hp and 8250rpm.
2.50	0.140	Extended short-circle-track and all-out drag race.
3.00	0.140	Long-distance endurance motors (Winston Cup, Grand National-type racing).

If you've no previous experience selecting pins for a particular application, this chart can be used as a reference for a good starting point.

Piston Rings Selection and Preparation

6

Piston rings perform a critical function: they seal the gap between the piston and cylinder wall. Without effective rings any engine would be a smoking, oil-burning unit delivering probably less than half its potential power.

Plain Top Rings

The most important feature of any piston ring is its cross section. The simplest type of ring, usually found in the top groove, is a plain ring. With such a name it sounds too simple, even lacking any real technology. But this is not the case. Although the plain ring consists of a simple cross sectional shape, it performs an important function.

The top ring must withstand and contain the high temperatures and combustion pressures of the combustion chamber. This means minimizing blowby and sealing against the high-pressure combustion loads seen around the TDC position. To do this, the ring seals as a result of gas pressure getting behind the ring and forcing it out, rather than by any radial pressure exerted by the static ring tension. A top ring must be made of a material that can withstand the shock of rapidly rising combustion pressure, yet retain its static ring tension at high temperatures, often on the order of 600deg. Fahrenheit.

Dykes Top Rings

In most applications the plain top ring can prove effective, but it isn't the only type that can be used in this position. One of two other types of rings, both of similar style, could be used. The first, the Dykes ring, is a lightweight ring named after its designer and intended for high-performance use. This ring consists of an L-shaped cross section, which utilizes the gas pressure to seal the ring against the bore. This ring type has many of the benefits of a gas-ported plain ring without necessarily having some of the high-wear disadvantages. Early in combustion, gas pressure builds behind the ring and quickly seats the ring against the bore and the wider ring face of a Dykes ring distributes the load better.

Headland Top Rings

The negative aspects of the volume contained above the piston ring in the top ring land were already discussed at length. It's a well-proven fact that any of the induced charge residing in the top ring land does not participate in the combustion event to any useful degree. A large ring land volume is known to reduce power potential and, especially important for modern engines, increase emissions. Recognizing this problem, Sealed Power developed the Headland ring.

In effect, the Headland is a large-section Dykes ring situated at the piston crown. It almost eliminates the dead volume normally seen in the top ring land area. My experience with this ring type shows it does an effective job with mean piston speeds up to 3,500 feet per minute (fpm). This relates to about 7000rpm for a 3.00in stroke motor, 6000 for a 3.48in as in a 350ci engine, and 5600 for a 3.75in stroke as in a 400ci.

Above these piston speeds the higher-than-normal weight of the ring can lead to reduced sealing capability due to the ring lifting from the bottom face of the groove. However, for good ring life combined with reduced emissions and a little extra power up to its rpm limits, the Headland ring has much to commend it.

Taper-Faced Second Rings

The second ring not only must assist the top ring in terms of compression, but also, it performs a *major* function in controlling oil consumption. When it comes to controlling oil, in most circumstances a plain-type ring falls short of optimum performance. To enhance overall oil control capability with the second ring it must have a form for scraping the oil down the cylinder wall. Many years ago, such a requirement led to the development of the taper-faced second ring.

The taper-faced ring has a tendency to ride over the oil on the upward stroke and scrape the oil off on the downward stroke. Various design improvements have been made to the simple taper-faced ring, such as a grooved lower face to enhance

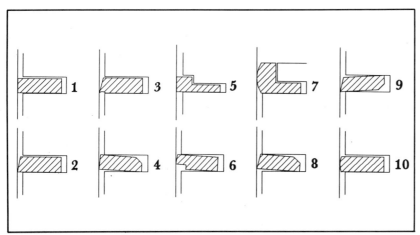

Here are the cross sections of ten common types of ring that can be found in the small-block Chevy engine since its inception in 1955. The Headland ring (7) has largely gone out of common usage.

The Dykes ring (5) is rarely used except on custom applications. Rings 1, 2, 4, 8, 9 and 10 have been used in literally millions of small-blocks.

the ring's ability to scrape the oil from the bore.

Torsional-Twist Second Rings

Probably the biggest step, though, toward improving the performance of the second ring is the torsional-twist ring. The original type of torsional-twist ring had a positive twist, that is, upward; this was achieved by machining a beveled edge on the upper inside rim of the ring. When such a bevel is machined on the ring, compressing the ring causes it to twist. When the piston and ring assembly is installed into the bore, the ring makes a line contact at the top and bottom of the groove. Combined with the line contact of the taper face on the cylinder wall, this improves the ring's ability to seal against both blowby and migration of oil up the bore.

The positive torsional-twist ring proved effective at sealing against blowby, and relatively effective at oil control. But because the seal was made on the inside edge of the lower face, some oil would get under the ring. At times of high pressures the ring would flatten out, causing some of the oil to migrate underneath the ring and eventually find its way into the combustion chamber.

Subsequent developments showed that if the ring was given a *reverse* torsional twist, it delivered superior oil control. With a reverse twist the lower face of the ring makes more contact and seals against the outer edge of the groove. Tests indicate that a reverse-twist ring gives better oil control under high-vacuum conditions than the positive-twist ring. A comparison between these two rings shows that the positive-twist ring has marginally better blowby control—an important factor for a high-output race engine—but the reverse-twist ring has better oil control, especially at high vacuum. This is an asset for a street motor.

Barrel-Faced Rings

The way the face of a ring presents itself to the bore can be important. We've discussed how close-fitting pistons, which exercise a greater control of the attitude of the ring to the bore, can have a *significant* effect on how well the piston-ring combination seals up. A certain amount of piston rock is almost inevitable, which means the attitude or angle of the

ring face against the bore will change. If a flat-faced ring is used, the ability to seal against the cylinder bore is directly affected by the angle the ring makes to the bore. On the other hand, a barrel-faced ring, which is a form

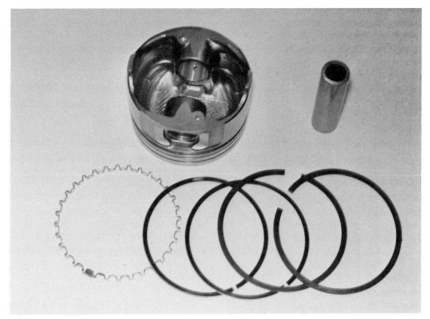

It makes no difference whether you're dealing with stock or race pistons—either way, care in gapping and prepping the rings is just as important.

becoming reasonably popular, has the advantage that even if the piston is slightly cocked in the bore, the ring is still likely to have effective line contact around the entire cylinder wall. The amount of barreling that a ring

The Sealed Power SS50 ring, as seen on this JE piston, is one of the most popular types of oil control rings on the market today.

needs to keep good line contact with the bore is typically only 0.0003-0.0004in.

Cast-Iron Rings

Originally, rings were made exclusively of cast iron because it contains free graphite, which acts as a dry lubricant giving a built-in antifriction system. On low-revving engines with relatively low cylinder pressures, cast-iron rings performed well. As a material for the top ring, however, cast iron had some shortcomings as engine rpm and cylinder pressures increased.

Cast iron is a brittle material, and high temperatures and the pounding brought about by high cylinder pressures cause cast-iron top rings to break. Since the conditions in the second ring groove are less severe, even today, cast iron is proving an effective material for the second ring.

Ductile-Iron Rings

For the top ring, something tougher was needed, leading to development of the ductile iron ring. In essence, ductile iron is a form of cast iron. But instead of the free graphite crystals being of random shape and array, the graphite is in a more controlled form, often spherical in shape.

If the iron contains a large percentage of free, spherical graphite, the type of iron is called spheroidal graphite iron. But in ductile iron the amount of free graphite is limited. By controlling the shape and amount of the graphite crystals the iron is made ductile, which means it is bendable and less prone to cracking. While ductile iron still presents good antifriction properties, its main asset is its resistance to cracking.

Hard-Chrome Rings

Though ductile iron is reasonably wear resistant, it's certainly not the best. In an effort to reduce wear, piston ring manufacturers have tried various techniques. Some years back a popular technique on more expensive rings was to hard-chrome-plate the rings as chromium has good antiwear properties.

Inspection of a micrograph of a chrome-plated surface will show what appear to be cracks or voids across its surface. These voids hold oil and therefore enhance the lubrication of the contacting surfaces. On a chrome-plated surface, 2 percent of the surface area is taken up by these voids. Since chromium in itself is a tough material, the overall effect is a ring that has extremely good wear life.

Molybdenum Rings

Probably the most popular ring type for the small-block Chevy, however, is the moly-faced ring, which has molybdenum metal applied to its outside diameter, and has a high void content somewhere in the neighborhood of 20 percent. Not only does molybdenum work well as a high-pressure bearing material, but its inherent hardness also means it wears slowly. Add to this the fact that it can hold oil in its surface, and we can see its potential for extended ring life.

Two types of molybdenum rings are available: the moly-filled ring where the OD of the ring has a groove cut into it and a molybdenum coating; and the moly plasma-sprayed ring. For the second type of ring the molybdenum coating is heated to virtually melting point and sprayed onto the OD of the ring. At such temperature it molecularly bonds with the ring's ductile iron. The advantages of the plasma-spray moly is that the moly is less likely to flake off.

Under normal circumstances the regular moly ring performs well, but if an engine is subjected to detonation, vibration and hammering of the ring can cause the moly to chip and flake off. If you are building a regular street motor the moly-filled ring will get the job done as well as any moly ring. But for race work where detonation may be a factor to contend with, the plasma-spray ring is preferable.

Ceramic Rings

Any material that is expected to produce a long wear life must have certain characteristics. One of these features is its reluctance to weld to any material, even on a microscopic level. One of the advantages to chrome and moly rings is they are not inclined to weld themselves to cast iron or steel. However, there are certain materials that cannot be welded, and ceramic is one of them. In the 1980s TRW introduced a ceramic-filled ring. In many ways ceramics share the same properties as a moly-filled ring. Ceramic has a porous surface and can hold oil, therefore, it should have good lubrication properties. But if you consider the fact that ceramics are virtually unweldable to cast iron, we see that the potential of a ceramic-faced ring is great. I've tested several sets of TRW ceramic-faced rings and though they have proved successful, the ceramic ring is still not as popular as the moly-faced ring.

Stainless-Steel Rings

For heavy-duty use, many engine builders opt for a stainless-steel ring. Stainless steel as a ring material has a number of advantages. First, the material is relatively impervious to high temperatures. Second, it is resistant to cracking and is highly

Here are three examples of different types of rings. The lower ring is a plain cast-iron type. The middle one is a plasma moly sprayed ring, while the top ring has a moly insert.

ductile. In addition, the high-chromium content of a stainless-steel ring displays many of the properties of a chrome-plated ring.

Under severe-duty applications, the chrome on a chrome-plated ring can wear off. Normally the plating is between 0.003-0.004in thick and although this represents considerable ring wear, we can find examples in off-road racing where similar wear could occur. Using a stainless ring, there is no problem with wearing off the working surfaces—it remains the same throughout the thickness of the ring.

However, stainless-steel rings, like chrome-plated rings, do take a considerable time to break in.

Ring Preparation: Gapping

Irrespective of the rings you select for your engine, it will be necessary to do some preparatory work before installation. If you are buying one of the more common brands of rings such as Sealed Power, TRW and so on, for a performance engine you should purchase the rings that require custom gapping to the bore size used. These are usually listed at 0.005in larger than the size on the bore; for instance, 0.030in oversize pistons will require the 0.035in rings.

In most engine rebuild shops the rings normally used are pre-gapped. For a normal street motor this is fine, as it saves time at assembly. But the ring gaps must be made a little on the large side so there is never a chance of the ring gap closing up entirely when the engine gets hot.

For a performance engine, though, pre-gapped rings are definitely *not* preferred. At this point you should consider time spent gapping the rings as time well spent. Keeping the ring gaps to a minimum helps power output measurably.

Prepping rings for your motor is not just a question of gapping them, however. To properly prep a ring you need some relatively fine crocus cloth, a fine lapping stone and a fine flat-needle file.

The first step is to gap the rings. Though time consuming, this is an easy job that most of us can do to near perfection by just taking the time and having the patience. Next, take the crocus cloth and remove any paint that may be on the outside diameter of the ring. If this is not done, the assigned ring gap will almost immediately enlarge by 0.002in as soon as the engine is turned a few hundred revolutions.

Once the OD of the ring has been cleaned it now can be located in the cylinder bore to ascertain the existing gap. With most rings requiring custom gapping, the existing gap will be less than zero so it will be necessary to remove some material from the end of the ring to establish a measurable gap.

To be able to measure the gap in the bore accurately, make sure the piston ring is square in the bore; if angled, the gap can measure out much larger than it actually is. The simplest way to ensure the ring is square in the bore is to pop a piston down into the cylinder, place the ring a good distance down the bore, then push the piston up until the ring squares up.

Ideally you need to check the ring gap about 0.750in down the cylinder bore. This may be a minor detail, but the reason for checking it farther down is that the greatest amount of distortion due to the head bolts takes place in the first 0.500in of the bore. Remember, we are gapping the rings without the benefit of the cylinder-head deck plate in place. Having removed the deck plate after the bore honing operation, the biggest out-of-roundness error occurs at the top of the bore. This error will not be compensated for until the cylinder head is in place.

If you want to speed up the process of locating the ring a given distance down the bore, then make a ring pusher using an old piston. Install an

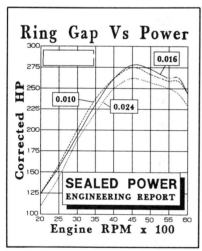

Closing down the ring gaps to a minimum value can make a difference in power, as this Sealed Power test indicates. When the ring gap was set at 0.010in, a better output was seen until the temperature of the ring reached such a point that the gap closed completely, and the increased friction caused the power to drop.

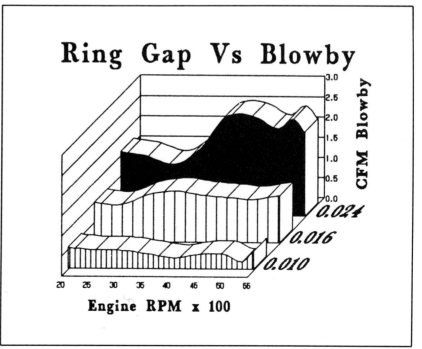

Here's the measured blowby as seen from the ring gap versus power test. Note how the reduced end gap considerably cut blowby.

oil ring assembly into this piston and use it to push the rings down the bore. When the oil ring butts against the top of the block, the rings you are gapping will be square in the block.

Adjusting Ring Gap

The traditional method of gapping a ring is to file the end of the ring. To do this you will need some soft jaws on a vise with a delicate enough touch so as not to break the ring, along with a reasonably fine-toothed file. To enlarge the gap you must file from the outside edge of the ring inward. Though not critical on single-material ductile-iron rings, it does become crucial on rings with any hard facing such as a typical moly ring. Drawing the file back over the material can cause the moly to chip at the edge.

Though filing gets the job done, gapping 0.0625in rings in this fashion is enough to try the patience of a saint, so let's look at some alternatives. The simplest ring-gapping tool is the item offered by Sealed Power. Its function is self-evident: you put the ring to be gapped on the tool's flat-top plate and crank the handle that turns the abrasive wheel. This takes material off either side of the ring gap. Though a bit cumbersome, it does beat filing by a big margin.

There are other alternatives as well. If you have an offhand grinder, with a steady hand and a good eye you can gap the rings quickly by grinding the end of the ring on the side of the wheel. Because it removes metal fast this technique requires that you exercise caution and good judgment as to how much material has to come off. Holding the ring on the side of a wheel—even a fine one—for just a few seconds too long makes the gap appreciably too large. From start to finish, using the grinding wheel technique is at least three times faster than filing.

If your budget can tolerate it, the ABS ring gapper is the ultimate tool for the job. A precision square gap can be generated, and once a measurable gap exists a precise amount can be removed so the desired gap can, within close limits, be achieved every time. Not only is this fixture far faster than any other technique, but it also offers far higher precision.

Amount of Ring Gap

The amount of ring gap you want is also crucial. Basically, the gap is present to compensate for the expansion of the ring as the engine warms up. Obviously the hotter the piston assembly runs, the more the ring will expand. Heat from combustion is dumped through the face of the ring against the cylinder bore and through the bottom face against the piston groove. The greater the amount of heat the ring can dissipate through these surfaces, the less it

will expand. Combustion factors obviously affect the amount of heat put into the ring, and a high-compression or supercharged engine will put much more heat into a ring than a low-compression engine. All these factors influence what the initial gap should be.

As a starting point you should use the manufacturer's gap recommendations, but if you are building a certain type of engine on a regular basis it pays to experiment a little. The gaps should slowly be decreased in size until the first signs of butting occur. My experience indicates that if rings are gapped in a precise manner it is possible to go 0.001in under that recommended by the ring manufacturer. Before doing so, though, be sure that the engine specs and the expected engine conditions will not subject the ring to any abnormally high temperatures.

Gapping the ring down to a minimum is fine for horsepower, but in doing so you may use up a safety margin. Depending upon the application, you need to consider whether or not a safety margin should be built in for, say, increased temperatures due to a motor lean-out or some other factor beyond your control. When considering the minimum gaps, also consider what the consequences would be if less-normal but still-possible situations developed. Obviously, for a drag-race engine one can reduce things nearer the limit because circumstances are far less likely to change during a sub-ten-second dash down the strip. Conversely, a error margin is needed if the vehicle's radiator should get clogged with mud and cause abnormally high temperatures. For a circle-track motor, such a scenario is not unlikely.

Reducing Ring-to-Bore Friction

The next step toward finished prepped rings is an exercise in friction and wear reduction. Using a lapping stone, the first step is to *lightly* stone off the sharp edges of the ring left after the gapping operation. But before going on to the next operation I want to explain why.

Looking at a piston ring we can see that the machining on the outside diameter of the ring and the top and bottom faces has left sharp edges at each corner. Assuming a ring sits squarely in the bore, those sharp edges wouldn't interfere with any-

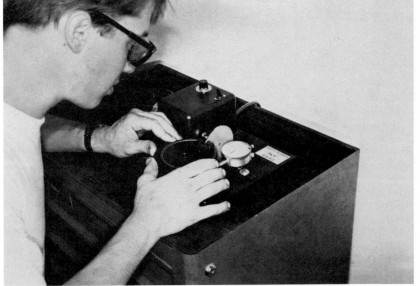

For the ultimate in accurate ring gapping, the ABS ring gapper is it. This is a must for all professional engine builders. *It allows fast, and more to the point, accurate ring gapping.*

thing. However, a combination of ring groove clearance and thermal and mechanical loads conspire to move the ring around in relation to the piston quite substantially.

Piston expansion can cause the outer edge of the ring land to tip down, cylinder pressures can cause it to tip up and inertia forces can cause both. On top of this we must consider that the ring is loose in the groove. There is usually about 0.002in ring-to-groove clearance. Under full-throttle conditions, gas pressures normally keep the ring on the bottom of the groove. During the induction stroke, however, the ring can move to the top of the groove. Going from one side of the groove to the other, the ring can bend. Be it ever so slight, this bending action first can put one sharp edge of the ring groove into the side of the bore, and then the other.

When a bore is freshly honed there are many minute peaks. Imagine these peaks moving into a sharp edge which is at a very slight angle to it and we can see the potential for the bore and ring to do a considerable amount of damage to each other. What we are dealing with is at microscopic levels, but the causes of most friction and wear are also at minute levels.

The real culprits in friction generation are the sharp edges on the rings. If the ring corners are stoned off leaving a radius of perhaps only 0.001-0.002in, a substantial reduction in bore-to-piston friction can be achieved. This friction reduction occurs only during the initial break-in. Even if the sharp edge is left on a ring, the ring and bore will wear until mating forms. However, we do not want to prep our bores by physically wearing them out in such a manner.

By just taking the sharp edge off the ring, the amount of time it takes the ring to bed in with the cylinder and the amount of wear that takes place is reduced. Also, the face of the ring seems to stay in better condition.

Yet if 0.001-0.002in radius helps, how would a bigger radius work? My experience indicates that once the edge is broken with the application of a small radius, an increase in radius achieves nothing. It may actually be counterproductive because if made large enough, it may counteract the gas-pressure-generated ring loads exerted against the cylinder wall on the power stroke.

It is all well to say that a radius of only a few thousandths inch is needed, but how do you measure a few thousandths inch? It is impractical under most circumstances and particularly this one. Basically, a few thousandths inch radius means the edge is no longer sharp. Try running your finger along the edge of a brand-new ring and it will almost certainly cut into your finger. Doing a lap or two around the edges of the ring with a fine stone removes this tendency, and this is how you will know if the radius is large enough.

When dealing with the second ring, it is important not to overdo the radius. This is especially true when dealing with taper-faced or Napier scraper rings, as they are sometimes known. The second ring in most ring sets has a taper face so it rides over the oil on the way up the bore and scrapes it off the bore on the way down. With this type of ring it is essential that moderation is used when applying the radius: two laps around with a stone is all that is needed. If you get overzealous with the radius, this scraper-type ring will not function as effectively.

Polishing the Rings

The last phase of ring preparation is to use crocus cloth to polish the rings, especially on the outer face. Again, too much enthusiasm should

be avoided. The idea is to polish the rings until they feel smooth—not to polish them so much that any surface coating is totally removed. The corners, however, will be subjected to more severe polishing so there will be the inevitable removal of any surface treatment right at the edges of the rings. When the rings are polished sufficiently it will be apparent to the touch.

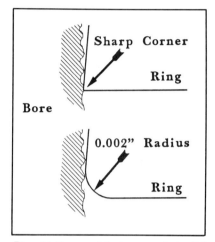

Ring friction contributes a substantial amount to frictional power loss. A 0.001-0.002in radius on the top or bottom of a ring can make a substantial difference to the rate of bore wear and frictional power losses in the early part of an engine's life.

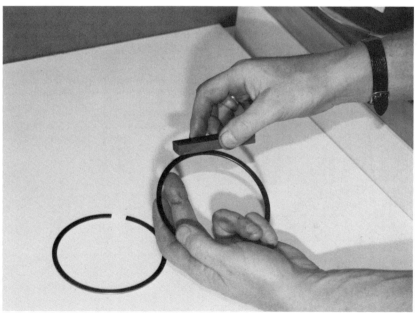

The first move for low-friction prepping of rings is to lightly stone the sharp corners off and then polish them with a crocus cloth.

67

Ring Preparation and Horsepower

By the time you have prepped a set of rings for your small-block Chevy in the fashion described, you will have a considerable amount of time invested in them. But how much is it worth?

Careful preparation such as this helps improve the engine's power output, it cuts the likelihood of any cylinder scoring during the break-in period, and it extends the useful life of the rings. The last factor is important because for most race engines, power drops off due to ring seal degradation. Since this is a primary cause of performance loss from a high-output engine, it is something worth dealing with.

As far as increased power is concerned, any that occurs is mostly brought about by reduced friction. Although friction levels ultimately reach the same point whether the rings are prepped this way or not, the friction level is lower earlier on in the engine's life. This is especially true in the first few minutes of running. Ring prepping such as this means your fresh engine can be run harder sooner and longer.

A simple check will show how effective ring prepping is at cutting friction. Just load up some gapped rings onto a piston and put the piston and rod assembly upside down into a newly finished bore. Turn the block upside down and hang increasing amounts of weights on the big-end rod until the piston just starts to slide down the bore. Next, remove the piston, prep the rings and try the test again. You will be amazed at how much less weight it takes to move that piston. Depending upon the bore finish and the pressures involved, the amount of weight can drop to as much as half.

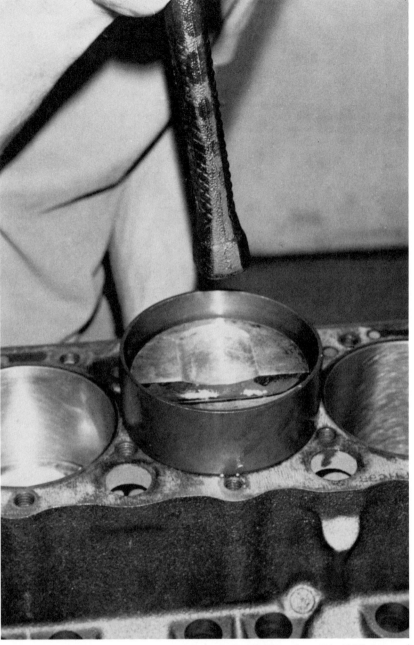

If you want to do a really fast job with the minimum chance of damaging the rings on installation, then this B&B tapered piston installer does a slick job.

Chevrolet Connecting Rods Selection and Preparation

7

With the exception of the 400ci connecting rod, all rods for small-block Chevrolets have a 5.70in center-to-center distance. But this does not make all rods equal.

Two predominant changes were made to Chevrolet rods over time. One entailed enlarging the big-end bore to suit the later big-journal crankshafts that came along from 1968 onward. The other change, which happened in two phases, was the beefing-up of the big-end bore. The 327ci motor was the first to see the squared cap; when introduced, the big-journal 350 connecting rod also had the top end of the journal bore squared. This squaring off of the big-end journal strengthened the rod so as to reduce ovalizing under more stressful operating conditions.

Unless you are building a small-journal-crank engine, you should always select the fully squared-off connecting rod as these are the strongest. Of course if you must stay with a stock connecting rod either because of budget restrictions or race rules, then the selection of the connecting rod needs to go beyond just choosing a rod of the correct pattern.

In its stock form, a small-block Chevy rod *tolerates* 7,000lb tensile load and delivers a reasonable rate of survival in a typical road-race application. With something around the standard pin and piston weight, this load capability represents approximately 3,200g reversal loads over TDC and these loads escalate rapidly as rpm rises. Though the rod can stand up to about 3,800g for short periods such as a few drag races, it's right on the ragged edge. To put this into perspective, there are plenty of aftermarket high-performance connecting rods that will run at 4,300g all day.

Looking at the situation in terms of how many g-force accelerations a rod and piston assembly pulls over TDC may be, for many, an unfamiliar way of viewing things. Let's break this down into rpm. What we basically have in stock form is a rod that is good for about 6500rpm. Try pushing that

to 7000 not just for a moment or two but on a continuous basis, and breakages *will* occur.

Now consider this: Most race-prepped small-block Chevys, especially the low-budget variety, are likely to produce their peak power around 7000-7300rpm. In the case of a two-barrel restricted motor, peak power will occur 500-800rpm lower, which for the present situation is good. If peak power does occur at the 7000rpm mark and the engine is required to overrev to 7500rpm, then reciprocating loads have just outpaced the stock rod's capabilities.

But all is not lost. With some fine-tuning, the level at which the stock rod can be used can be raised significantly, though of course it will never match a purpose-built race rod. The fine-tuning procedure begins with the selection of a usable rod set and then continues with the preparation.

Rod Selection

If circumstances dictate the use of stock rods, then the prime require-

ment is to find eight *good* rods. If selecting eight rods proves too costly and the old ones must suffice, skip this section and move on to prepping the rods. However, if the opportunity arises to delve through some friendly motor machine shop core

Rods do break. If all other areas are properly prepared, a breakage like this is about the most likely.

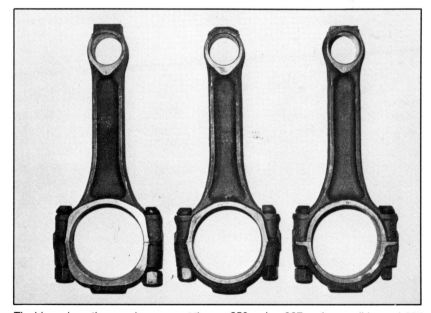

The big ends on these rods represent the three different styles of rods you can get. From left to right, they are a big-journal 350 rod, a 327 and a small-journal 265 rod.

bins, then some interesting avenues open to you.

First, in order to get eight good rods you are likely to need ten rods as closely matched as possible to select from. Considering the kinds of features needed, it will probably require a core bin with as many as 100 connecting rods in it. So figure it could take you awhile.

When confronted with a pile of rods, the first thing to do is avoid selecting rods that have anything but the mildest surface rust. Rust will pit the surface, and this will be the first place to propagate a crack. Rusty rods will break sooner than others, and the fact that you may be able to bead blast the rust does not give the rod a clean bill of health. Rust can pit well below a level at which a bead blaster can clean and if it's in a critical stress area, this is where a failure will originate.

Rod Balance Pads

The next step is to look for rods with the *smallest* balance pads at either end. This may seem like a useless task. Surely if the balance pads are as big as possible there would be more metal to remove and you will end up with as light a rod as possible. This may be true, but the object is to select rods for strength rather than weight because a marginal-strength situation exists.

If a rod has small balance pads, then, since all factory rods are supposed to weigh the same, the metal must be somewhere else in the rod. If it's somewhere else it's probably doing some good because the only really useless material in a stock rod resides in the balance pads. Other than to standardize the weight, the pads are just along for the ride.

When selecting connecting rods based on the balance pad size it is better to select ten rods that are, as far as the eye can tell, identical rather than trying to get ten rods that represent the ten smallest balance pads in any group. In other words, if one connecting rod has virtually no balance pads to speak of and seven more have a small balance pad, then the zero-pad rod is going to be an orphan. You don't want that rod in a group of significantly different rods because it could end up providing a balancing problem later on. Though minimum balance pad size is preferable, try and select rods that match as closely as possible.

Matched Rod Dies

Phase two of rod selection involves determining how well the dies matched during the forging process. If you check out a few rods, you will see that for the most part the dies don't exactly match and one half of a rod is slightly offset to the other; this is especially noticeable in the beam. In going through the rod pile, chances are you will eliminate at least 30 percent.

From the satisfactory die alignment rod group the next inspection

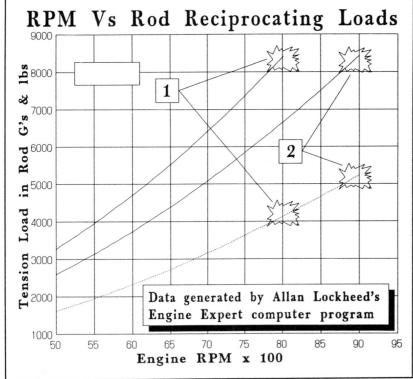

RPM Vs Rod Reciprocating Loads

Data generated by Allan Lockheed's Engine Expert computer program

Tension Load in Rod G's & lbs (vertical axis, 1000 to 9000)

Engine RPM x 100 (horizontal axis, 50 to 95)

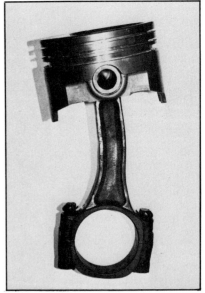

The pink rod, which is Chevy's heavy-duty rod, comes out of the same dies as the stock rod, but is made from a tougher material, and is reckoned to be about 20 percent stronger. This is another bent rod from the turbo engine. This Chevy pink rod came out of a turbo motor that used about 30lb of boost. Cylinder pressures were so high that all eight rods bent like this.

The figures from this graph should be used to indicate a trend, rather than absolutes. If we make the assumption that a typical stock rod, if run to 8000rpm for only a matter of seconds, will break, then, by fully preparing the rod and hanging the lightest possible piston on it, we can increase the breakage point by almost 1000rpm. The solid upper curve represents the load in the rod with a stock weight of piston. The rod breaks at about 8,300lb. The middle curve represents the load on the rod when a lightweight piston and pin assembly is used. Note the rod still breaks at about 8,400lb, but this loading occurs at virtually 1000rpm higher. The lower dotted line represents the piston imposed G forces seen at the pin in either case.

point will be the small end. It's not uncommon for the small end to be bored quite a bit off center, leaving one side of the small end thinner and the other side thicker. A rod noticeably offset must be rejected. This usually means transferring yet another 30-50 percent of the rods to the core bin.

Reconditioned Rods

Even though we're nearly to the end of our selection routine, you will still need more from the eight required rods because some of the rods you've chosen may have been reconditioned. Although this may not seem to be a problem, as a rule it is.

When reconditioning a rod many engine rebuilders are concerned with how the finished rod looks to the customer. To recondition the big end, the caps are normally ground down a couple thousandths inch to reduce the size of the big-end bore, then it is re-honed to size.

Grinding the caps down does not bring the sides of the rod in at the split line, however, and this dimension will stay virtually the same. In most cases the honing operation does

not clean up the rod journal bore at the split lines. Usually a patch of the old finish is left, giving the rod a less than perfect appearance, though functionally it can have a clean bill of health.

A technique used to avoid the unfinished patch calls for angle grinding the caps. This technique involves grinding each side of the cap at a small angle; likewise on the rod. The grinding is done such that the inside edge nearest the bore at the split line is high so that as the cap is tightened to the rod it tends to pull the sides of the cap in. Any subsequent honing operation cleans the rod journal bore thoroughly because the sides of the rod are pulled in.

This in turn bends the rod bolt, the outside of the bend being nearest the rod journal bore. At high rpm, the rod bolt will bend this way as the rod journal bore elongates under stress.

A rod bolt should *never* be subjected to bending loads. The fact that a small-block Chevy rod bolt *does* experience bending loads means it will fail earlier.

Usually a bolt failure is the most common type of failure for the small-block Chevy rod. To understand why, here is a scenario of what happens. First we need to realize that material outboard of the bolt at the split line tends to support the cap from ovalizing. It's there as a kind of lever. As forces try to ovalize the rod journal bore, pressure is applied to the material outside of the rod bolt and this in turn acts as a backstop to prevent ovalization.

A small-block Chevy rod has too little material outboard of the rod bolt, so its tendency to bend the rod bolt is higher than would otherwise be the case. By angle cutting the cap a certain amount of bend is put into

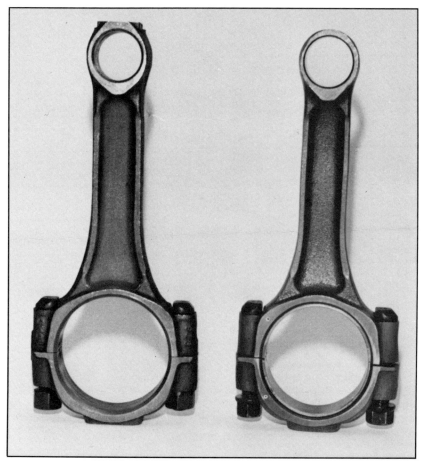

This is a rod little end that has been bored too much off center, and has left the material thin on one side. A prime candidate for a failure.

Check out these two rods: the one on the right had a large balance pad before it was removed. Notice how the beam section of this rod is less than the rod on the left, which has a small balance pad. Hav- *ing a small balance pad is a good indication that the material is elsewhere in the rod. In this instance, it's in the beam where it can contribute to extra strength.*

the rod bolt to start with. This lessens its capability to take further bending. The result is early failure of the rod bolt if the cap has been angle cut.

The question is, without some precision equipment, how do you know if a rod has been reconditioned? The first step is to check the finish on the split line face. The split line face on most factory rods is machined somewhat coarsely because it was probably done on something like a multitoothed milling machine or segmental grinder. On the other hand, most rod reconditioners use a Sunnen cap grinder, which puts a fine finish on the split line surface. A fine finish generally indicates that the rod has been reconditioned.

There is the possibility that the rod has been reconditioned by a rebuilder who didn't employ the angle grinding technique, in which case the rod might be OK. However, it's still a dubious choice to stick with any rod that has been reconditioned. You can look at it this way: if a rod has been reconditioned, it's probably seen high mileage and much of its fatigue life may have already been used up. Although a rod that has *not* been reconditioned has no guarantee of a long fatigue life, its chances are certainly better than those of a reconditioned rod.

Prepping Rods: Mechanical Reliability

Essentially there are three aspects to preparing a connecting rod: modifying to improve mechanical reliability, lightening the rod, and precision-fitting all critical dimensions. To an extent these factors overlap, but where possible we'll deal with each individually.

Improving reliability is a major issue with the stock rod. Rod bolt failure is probably uppermost in people's minds when dealing with stock rods, but this is a problem that can be addressed relatively easily by installing heavy-duty rod bolts. We'll cover that subject later. For now we'll deal with the actual rod forging and its inherent weaknesses.

The first of these is the rod bolt notch in the top half of the rod. The red bolt head butts up against the side of the notch to prevent the bolt from turning while the nut is being tightened down. Unfortunately, the position of this notch produces a critical stress area in the rod. The notch acts as a stress point, and flex at this point will cause the rod to break. The question is, what can be done about it?

Usually if something is not strong enough at a particular point, all that can be done is to add metal to it for strength. Although this would remedy this particular situation, another partial remedy can be effected by *removing metal*. By increasing the radius or enlarging the flex area, the tendency for fatigue fracture at the notch is reduced.

I cannot overemphasize the importance of reducing stress concentrations. Fatigue acts primarily on stress points and brings about early failures. Without actually making the material thicker or stronger, the fatigue resistance is increased.

This same technique can be applied to a connecting rod. Most connecting rods have a reasonable radius at the notch, yet some may have quite a sharp angle. If any of the

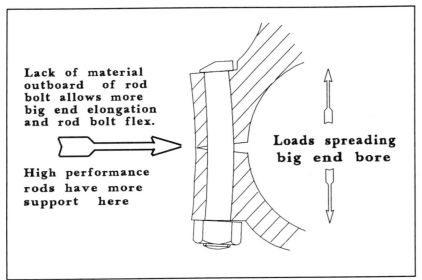

The effects of high reciprocating loads on bending the rod bolt lead to early breakage. The outside of the curve becomes overstressed much sooner than the material down the center line of the bolt. This usually leads to early failure, usually around the split line level.

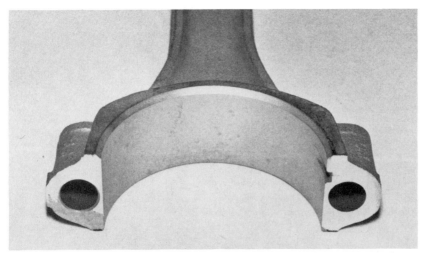

The split line of this rod has been ground on a Sunnen grinder, but I've stopped just short of cleaning up. Notice how some of the original surface remains on the left-hand side of the rod. This indicates that this face was not machined square to start with—just one more reason why careful reconditioning of the rods can pay off in reliability in a high-performance engine.

selected rods have tight radius notches, get rid of them—they're trouble waiting to happen.

The need for the notch to stop the bolt turning is not super-critical. If a good-quality rod bolt is used and the bolt is pressed in such that the edge taking the turning torque is against the face, then the friction level of the bolt in its locating hole is usually enough to prevent rotation as it is tightened. And if Loc-Tite is used on the bolt, the need for the notch is virtually eliminated.

If this is kept in mind, we can see that making the radius in the corner of the bolt head notch as large as possible is not going to have any negative side effects on bolt retention. But making it larger will improve the fatigue strength at that point quite substantially.

If the bolt head face is not square to the bolt hole, tightening the rod assembly will produce bending loads just under the bolt head. To check the squareness of the bolt hole and face, machine up a bolt on the shank and the underhead surface. The shank needs to be a snug fit in the bolt hole. Using a *light smear* of engineer's blue on the head of the bolt, install it and turn it from side to side. Then remove it and check to see how the rod marks up. The blue will show high spots, which can be corrected with a needle file.

Rod Lightening

Moving up the flanks of the rod, de-flash the flanks with a grindstone. Work *lengthwise* down the rod so imperfections don't form tiny notch grooves across the rod. Anytime ma-

terial is removed from a connecting rod, the balance will be affected. Taking it off the flank is the first step. Similar lightening treatments will be applied to both the cap and the small-end balancing pads.

The question is how to keep the rods of similar weight while lightening them all. For this I use the APT balancing fixture. This relatively inexpensive fixture allows rods to be balanced to a finer precision than 95 percent of professional rod balancers. It also allows you to keep track of rod-to-rod balance as you lighten the rods.

The technique when using the APT balancer is to rework one rod at a time, starting with the rod you expect to end up the lightest. Having completely reworked the one rod, all the other rods should be progressively reworked to match it.

Returning to the cleaning of the flanks, after having rough ground them and removed any obvious notches from the beam of the rod, you need to dress the flanks with a relatively fine emery drum. Again, work lengthwise on the rod, not across it.

Precision Fitting

The next subject on the agenda is the rod small end. The hot topic here is how the wrist-pin-to-rod assembly

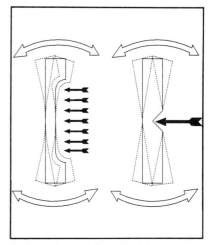

A piece of material notched down to give the same minimum thickness as a piece carefully cut away with radius ends will fatigue crack much sooner, because the stress is concentrated at one point. In the left test piece, the stress is spread over an appreciable area. This illustrates why removing material around the area of the bolt notch increases the fatigue resistance of the rod.

<hr>

Shot Peening

Shot peening is probably one of the most cost-effective, readily available surface treatments to improve the life expectancy of crankshafts and connecting rods. All too often, though, shot peening is confused with shot blasting or glass beading.

Shot blasting, be it with sand, steel grit or carbide shot, does not achieve the same effect as shot peening. Any process that uses sharp-edged particles will almost certainly *reduce* the strength of the component you're trying to improve. Therefore, do not contemplate any of these processes as a substitute for genuine shot peening.

When a component is shot peened, smooth, round steel shot is impacted on the surface being treated. Each shot ball leaves a smooth, spherical impression, and compacts the skin of the part, causing a compressive surface stress. Since parts break from tensile loads opening up minute surface flaws, we find that the compressed surface layer delays the onset of failure. When done correctly, not only will the part take as much as 20 percent more in terms of maximum load, but its fatigue life also may be increased 20-50 percent.

To have shot peening done, I recommend using the services of a company that specializes in the field. Look through the Yellow Pages for companies advertising shot peening. It's best to work with a company that is capable of shot peening to military specifications MIL-S-13165C (or a spec that

supercedes that). Usually any company that is dealing with aerospace components can handle this. Also, you may find that the companies which perform aircraft maintenance may have shot-peening facilities on hand, as most aircraft crankshafts and rods are shot peened. You can also phone the executive offices of the Metal Improvement Company, a subsidiary of the Curtis Wright Corporation. It has shot-peening branches all over the United States.

Once a specialist is selected, if that company deals with many components other than crankshafts and connecting rods, they may want to know what specification the parts should be shot peened to. Talk with whoever will be doing the work. Explain the application thoroughly to ensure the best job possible. When sending in the parts, be sure that any masking leaves the most critical areas exposed to the shot. This means making sure the fillet radii can be more than adequately shot peened, and the rod bolt head notches of the rods.

It also means installing temporary bolts that are cut away to allow the shot into the corner of the rod bolt shoulder. Remember, if the rods are prepped as has been detailed earlier, then you should have polished this area and even enlarged its radius. The icing on the cake is to make sure the shot peening performs its important function.

The first move to increase the reliability of your rods is to enlarge and polish the rod bolt radius.

structure. In so doing it relieves the pistons of having to hold a pin that is flexing; flexing pins can often be responsible for cracks in and around the piston bosses. All other things being equal, a piston utilizing a fully floating pin requires slightly more strength in the piston boss to survive than one used with a press-fit rod. But beyond this minor advantage, a press-fit pin has some major disadvantages.

First of all, the press-fit pin is much more difficult to assemble. This will usually entail heating the rods, which would be a no-no if any heat treatment or shot peening had been done on them. A second disadvantage is that removing the pistons from the rod almost inevitably destroys the piston. On the other hand, when putting together an engine that you expect to last 100,000 miles, a press-fit pin isn't such a bad deal—especially if the rpm limitations of the engine do not warrant any special rod preparation.

When it's time to rework the small end, start by grinding off the balance pad. If you don't have the means to balance the rods, bear in mind you will need to make a good estimate of how much is removed from each rod for consistency. If you have an APT rod balancer, then achieving the balance is no problem, it's just a case of reworking the rod down to the desired weight.

should be treated. In stock form the wrist pin is a press fit in the rod small end. Many racers, and certainly most aftermarket rods, employ a pin that is free to turn in the rod, thus making it a fully floating small end. So what are the advantages to a fully floating small end? There are pros and cons.

There is an advantage to retaining the press-fit small end, in as much as this strengthens the pin as a beam

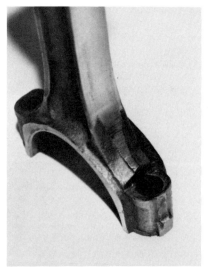

This is typically how the area just above the bolt hole looks prior to reworking. Don't be concerned about excessive metal removal here.

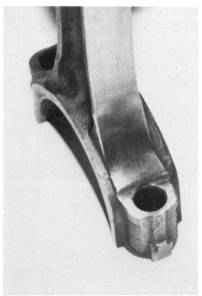

Thinning the bolt notch down to the form shown here actually increases the rod's fatigue resistance at this point.

When grinding the flank of the rod, be sure to grind lengthwise, as this is less likely to produce crack propagating stress risers.

At this point the obvious question arises as to how thin the wall section can be around the small end before failure is likely to occur. Any lightening done to the rod depends on the treatment the small end receives. If the small end is made to be free-floating with a bronze bush, then obviously the material around the small-end bush must be thicker to start with. This is because some of the strength is taken away by bushing the rod initially.

If the rod is to retain its press fit, then the section thickness around this point can be decreased a little more since no bushing thickness allowance needs to be made. Typically a bushing for the small end has a wall thickness of about 0.025in, so this will entail boring the small end 0.050in oversize to accommodate the bush. When lightening the small end for a bushed rod the extent of the lightening operation must reflect this requirement.

There is a practice among engine builders to simply hone out the small end and make it fully floating with a steel-on-steel bearing. This technique produces mixed results. If the combination of power and compression ratio aren't too high it seems to work adequately if the right lubricants are used in the oil; Crane Break-In Lube added to the oil seems to work well.

As compression ratios increase, pin wear accelerates. For example, whereas a 9.0:1, two-barrel circle-track motor may run an entire season with a steel-on-steel floating pin with no measurable wear, a 14.5:1 engine with an hour's full-throttle running can sometimes put as much as 0.001in wear on the pin. In spite of its greater hardness, it is usually the pin that wears rather than the rod. Though a steel-on-steel floating pin may be up to the job for a drag-race engine or a restricted low-compression circle-track motor, such a technique is not recommended for the street or even a semi-serious race effort.

The problem of steel-on-steel bearing wear can, to a large extent, be eliminated with surface treatments. For instance, when Tuftriding—a process whereby steel is heated at a high temperature in an atmosphere of molten cyanide—was in its heyday before pollution regulations shut down most plants, a rod treated by this process showed a considerably extended life in terms of pin wear. These days, viable substitutes are nitriding, Nitemper and a few other heat-treatment processes.

When it comes to rebushing the small end, most engine rebuild shops have the equipment to do the job. The bushing operation also allows the center-to-center distance to be set to close limits. A consistent center-to-center rod length removes one variable when it comes to matching compression heights at assembly time.

As far as sizing the pin bushing goes, this needs to be done in relation to the pins you're using as one brand of piston varies in size to another. Also, the clearance required may vary significantly from one application to another. If the engine is to be used for drag racing where the pin end of the rod has too little time to get up to working temperature, then clearances need to be a little larger. On the other hand, for a street application where the engine's likely to be

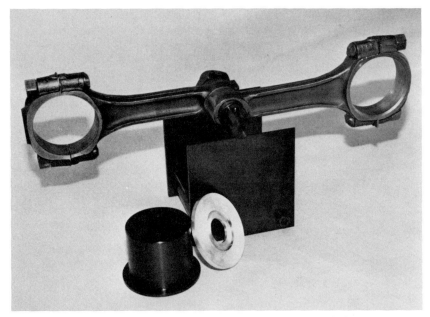

The most difficult problem with lightening your own rods is making sure they can be balanced. This balance fixture is used in this manner to balance the big ends.

The balance fixture is then reversed like this to balance the little end. This allows you to rework your rods evenly down to a minimum weight, leaving them completely balanced at the finish. The accuracy of this system is two or more times that of the typical balance scales used in most engine shops.

warmed up thoroughly before being used at high rpm, clearances can be tighter to avoid any small-end rattle. For most applications a snug slide fit of 0.0002-0.0003in clearance will work just fine. For drag-race applications this needs to be up to about 0.0005in and if nitrous oxide injection is used, clearances need to be enlarged to 0.0006-0.0007in.

Once a small-end bush has been installed there is the question of lubrication. The usual technique is to drill a hole in either the top of the rod,

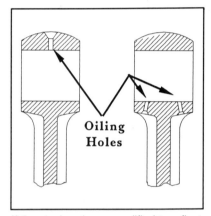

If the stock rods are modified to a floating-pin style, then some form of pin oiling will be found to help the life of the little end. Two commonly used techniques are shown here—I don't particularly recommend either, as they can lead to breakages. The text explains how to lubricate the pin and retain maximum reliability.

Here is the application of the technique we used to preload the big-end bore so that it would produce a round bore at an estimated 3000-4000rpm.

or two angled holes in from the flank. Although both techniques can actually cause a breakage at the small end, the two-hole method is the least reliable.

Probably the best method is to simply file an oiling groove 0.010-0.015in wide and deep across the width of the bearing. This can be at the top or bottom of the little end without affecting the bearing load capability. However, side grooves work just as well and don't decrease bearing area in a highly loaded spot.

Rod Bolts and Big Ends

The stock rod bolt—which, if used in pure tension, should be strong enough for most applications—will fail much sooner than it should. We discussed the technique some engine builders use for angle cutting the cap, which bends the bolt. But this bolt bending is exaggerated even further as rpm increases. The bending is normally kept in check in a well-designed rod by having sufficient metal on the outboard side of the rod bolt to support the rod, reducing ovalizing.

With a stock small-block Chevy rod this is not the case; as rpm increases, the sides of the rod come in, bending the bolt and overstressing it on the inner side of the shank. What happens is that the material on the inner side will now fail at any slight imperfection. Once it starts to fail, the weakness quickly moves through the rod bolt and the bolt breaks.

The sides of the rod journal moving inward can also affect the oil film. Bearings typically are relieved at the split line, so the clearance is larger there than anywhere else. Thus as a rod collapses at the sides at high rpm, the bearing itself does not end up scraping the oil film off the bearing.

To get the best from the big end of a stock connecting rod we can prepare it the easy way, or we can get a little more from the rod in terms of reliability at high rpm by prepping it the hard way. First, the easy way.

Basic Big End Preparation

The initial step is to flat grind each half of the rod and cap assembly, as opposed to angle grinding. Each half then needs to be deburred around the edges and at the hole. The rod and cap should now be bolted together, torqued up and the rod journal bore sized.

This is all straightforward enough in itself, but two factors need to be taken into account. First, on a small-block Chevy running high compression, the timing doesn't need to be too far out or the fuel octane just a little off to run into detonation, which tends to cause spun bearings. It could well be that the detonation itself hasn't caused any problems on the piston or combustion chamber, but inadequate bearing retention and its subsequent spinning in the rod has now caused the rod's demise. If not caught instantly, the crank will quickly follow suit.

The way to reduce this possibility is to make sure the bearing retention is up to snuff. The tangs on the bearing are not there to retain the bearing, as is often thought: they simply locate it into position. It is bearing crush in the rod that holds the bearing in place.

When getting rods reconditioned, be sure that the person running the rod-honing machine understands that this is for a high-performance application and sizes the rods toward bottom limit for maximum bearing crush. Also, the bore finish for a stock rod should be accomplished with a coarse stone; the coarser finish holds the bearing better than a fine finish.

To ensure that the bearing retains its crush, even after a severe pounding, it's a good idea not to chamfer the split line in the journal bore other than simply to remove the sharp edge. A large chamfer here can actually allow the bearing crush to ease over time.

Performance Big End Preparation

Now let's look at a more complex method, which is the way to go if big-end reliability is a major concern. As mentioned, the problem with the rod is insufficient material outboard of the rod bolt, allowing the rod journal to ovalize easier. We can combat this to an extent by producing a rod bore that is oval across the other direction so that under stress, its roundness increases in the vertical axis.

To do this, the rod caps are angle ground the opposite way engine rebuilders normally grind them. The result is that the cap or rod is low at the journals side so that the action of tightening a rod bolt causes the rod to spread. This bends the rod bolt outward—the opposite from the way it bends in service. Once the rod is torqued, clamp the sides of the rod to bend the rod back in allowing, with a certain amount of pre-load, a round journal to be produced.

This process is similar to using torque plates for honing cylinder heads: essentially we're prestressing everything in the opposite direction so that in use, a certain rpm must be reached before the rod bolts begin to bend inward. Thus for a given rpm above a certain point in the rpm range, the peak surface stress seen by the rod bolt is reduced.

Unfortunately, this technique is time consuming and requires fairly accurate machining. First, the caps must be angled only 0.0005in across the face. This produces a bore from 0.0005-0.00075in out-of-round, being too wide by this amount at the split line. By clamping across the width of the rod and measuring the bore size change we can pre-load the bore inward about 0.0005in to a maximum of 0.00075in. Having been pre-loaded, the bore is then honed to size.

It's difficult to say what this technique is worth in terms of extra reliability. Over a period of time I've prepped about half a dozen sets of rods in this manner—too few to use as a reliable statistical base. But even this limited experience reveals an improvement as far as rod bolts are concerned. Some of the rods were equipped with stock bolts and went the distance under conditions in which they were normally expected to fail.

Once your connecting rods are fully machined, lightened and generally prepped, there are several processes to consider to increase the fatigue resistance of the rod. These processes, by one means or another, combat fatigue by producing a compressive stress in the surface. To understand why this improves a component's fatigue strength we need to appreciate the fact that failures al-

	1. Stock		2. Stock		3. ARP	
Torque	Load	Stretch	Load	Stretch	Load	Stretch
25lb-ft	5.75	Bad readings	5.75	—	4.25	2.0
35	7.3		8.0	4.0	5.5	3.8
45	8.75		9.2	5.0	6.6	4.0
50	10.4		11.0	6.8	7.1	4.1
55	14.2		11.75	7.2	7.2	4.5
60					7.5	5.0
75					11.8	10.0
90					14.0	
95					14.5	

Rod Bolt Test

Note: Stock bolt was well into yield at 75lb-ft. One went to 14,200lb, the other to 13,000lb. The ARP bolt took on a set of 0.0045in increase in length after 95lb-ft had been applied. Yield on the stock rod bolts was about 1½ turns while yield on the ARP bolt was mild.

Here are the results of some testing done in my shop on a rod bolt test machine. In terms of outright strength the figures obtained for the ARP bolts speak for themselves. The ARP bolts, however, take more torque for a given amount of clamping load.

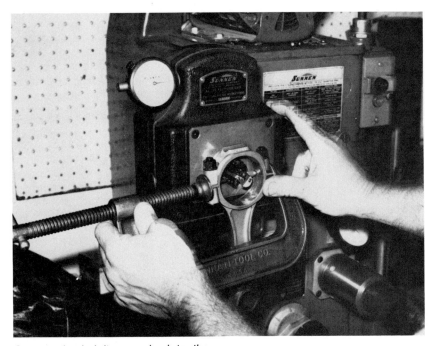

Once preloaded it was sized to the bottom limit to give maximum bearing crush.

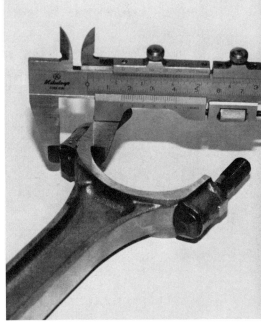

Building a 383 or 400 with a 5.70in rod? If so you will need to grind the rod as shown here to 0.0625in to clear the cam.

77

most always begin at or near the surface of a component, and that fractures are the result of tensile stresses acting on features that are known as stress raisers.

These stress raisers can take many forms: simple machining marks can be stress raisers, as can minute surface flaws caused by, say, almost undetectable rust particles. This and countless other surface flaws can be the source of the beginning of a crack. By processing a component so that it has a compressive layer in the surface, tensile cracks are harder to start and propagate. The effect of a compressive surface stress in a component such as a connecting rod can have a significant effect toward extending its fatigue life.

Performance Aftermarket Rod Bolts

The purpose of the rod bolt is to hold the cap to the rod with enough pre-load that even at peak accelerations over TDC the cap does not separate from the rod. It's not enough that it does this just a few times, but millions of times. The combination of the stresses involved, plus the repeated load reversals means the material of the rod bolt not only needs to be strong, but also tough to resist fatigue.

From this point of view, it's not enough to look just at a material's tensile specification to determine that the parent material is adequate. For instance, a stock Chevrolet bolt's tensile strength is on the order of 170,000-180,000lb per square inch, yet it is possible to make a rod bolt from a material with an identical tensile stress level that will outlive the stock rod bolt by a factor of two to three times. Because of the design of a small-block Chevy rod, it is ultimately fatigue that kills it rather than the strength of the material.

Tensile strength and fatigue resistance are two factors that can't be analyzed separately. For instance, a rod bolt that is able to take a higher stress level can exert a higher clamping load, and this in itself may keep some of the localized bending in check, thus cutting fatigue failures.

Another factor that must be considered is quality control, not only of the material involved, but also the end product. Remember, there is not much point in having fifteen excellent rod bolts in an engine and one questionable one; when that one fails,

Tensile Test Results on Five Randomly Sampled Bolts

Bolt	Original Equipment Bolt Tensile Strength (lb)	Proportional Limit (lb)	SPS/Mr. Gasket Bolt Tensile Strength (lb)	Proportional Limit (lb)
1	17,020	15,000	17,500	14,000
2	16,540	14,120	17,740	15,000
3	16,460	13,000	17,680	14,800
4	16,800	14,600	17,660	13,600
5	16,620	14,400	18,200	14,600
% Spread	3.4	15.4	4.0	10.3

Minimum Tensile Strength of Samples Tested
Stock: 187,960psi*
SPS/Mr. Gasket: 201,000psi*
*Based on a working diameter of 0.334in.

Notes: Tensile strength indicates how much load was required to break the bolt. Proportional limit is where the bolt starts stretching at a rate proportionally greater than the rate at which extra load is applied. This is a sign that the bolt is close to its breaking point.

Alternating Stress (lb-in)	Original Equipment Bolt Cycles to Failure	Failure Location	SPS/Mr. Gasket Bolt Cycles to Failure	Failure Location
± 42 Ksi	8,000	Thread	105,000	Thread
	10,000	Thread	105,000	Thread
			61,000	Thread
± 32 Ksi	17,000	Thread	166,000	Thread
	19,000	Thread	433,000	Thread
			290,000	Thread
± 28 Ksi	30,000	Thread	245,000	Thread
	28,000	Thread	303,000	Thread
± 20 Ksi	98,000	Thread	1,019,000	No Failure
	76,000	Thread		

Notes: Because the SPS/Mr. Gasket bolt is made of tougher material, note how much longer it takes to fatigue fracture. Generally speaking this bolt lasts about eight to ten times as long as the stock bolt.

The reliability of a bolt is not solely connected with its tensile strength; its fatigue life is also important. Just study these figures and you'll see that buying a high-performance bolt means buying a great deal of extra fatigue resistance as well as extra strength.

The stock rod bolt (right) failed at 75lb-ft, whereas the ARP bolt on the left only just started to fail at 95lb-ft. Greater strength and substantially greater toughness make these bolts desirable pieces to use in your small-block Chevy rods.

it takes out the whole engine. Because of critical service demands, rod bolts need to be inspected and replaced if there's any doubt.

When buying a quality aftermarket rod bolt, you're getting a higher level of inspection than with the stock rod bolt. Along with better material, this is the best insurance against bolt failure. However, what you do with the bolt ultimately dictates the kind of service life it delivers.

To begin with, a rod nut should always be torqued onto its bolt with a high-quality, high-pressure lubricant. Moly grease or Crane Cam Lube are suitable and recommended for this application, though there are many other satisfactory alternatives. Without a good lubricant much of the tightening torque is lost in overcoming friction, so the clamping forces exerted by the bolt are significantly reduced.

Tightening Rod Bolts: Torque Versus Stretch

To deliver the correct clamping force, ARP recommends that the bolt be tightened down to deliver 0.006in stretch rather than quoting a torque value. Torquing a bolt only implies the right amount of stretch, and thus preload, but it is affected by the level of thread friction existing. Not lubricating the threads properly can drop the amount of clamping force for a given torque by 2,000lb and as much as 3,000lb.

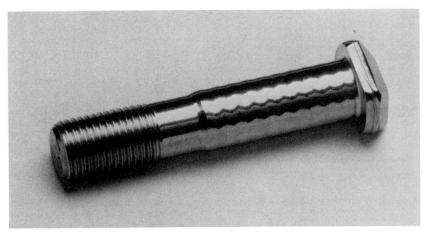

Here's ARP's latest rod bolt as of 1992. Instead of knurling the press-fit part of the shank it is machined with this wave form. This makes for a stronger bolt and allows bolt changes to be made, providing it has already been reconditioned with a wave-form bolt installed.

ARP bolts tended not to deliver as high a clamping load per lb-ft of torque as the stock ones. However, the amount of clamping force delivered is more than enough to keep the two halves of the rod from separating, so no additional tensile load is put on the bolt. In fact, if it were strong enough, a stock rod and piston assembly would have to be turned to 11,300rpm to experience any separation with the *stock* rod bolt.

Like the ARP items, the SPS bolts as sold by Mr. Gasket scored well in terms of fatigue resistance. It's ultimately this characteristic of a high-performance bolt and not its peak tensile strength or clamping load capability that dictates how long it will live.

When it comes to tightening high-performance bolts a rod-bolt stretch gauge should be used, since the amount of stretch is what determines the bolt pre-load, not the torque. For those not so equipped, which is most of us, ARP recommends the following technique for a 3/8in Chevrolet rod-bolt assembly. Using moly grease on the threads and the nut face, torque the assembly to 50lb-ft, then back it off completely. Regrease the nut and retorque again to 50lb-ft. Repeat this a third and final time.

Aftermarket Connecting Rods

8

Until now, we've dealt with stock factory connecting rods and unless it's a 400ci engine you are working with, there is no choice as far as center-to-center length is concerned. Because of this, the main issues we have focused on are weight and strength. Let's assume finances allow the purchase of one of the numerous types of aftermarket rods available. This poses many questions, of which the single most important is whether or not the stock rod length should be retained. If not, what are the benefits of changing rod length? Understanding rod and crankshaft geometry and its effect on the engine can get complicated, but the potential benefits make the effort worthwhile.

Rod and Crankshaft Geometry

When dealing with crankshaft strokes and rod lengths it is proportions that count, and to define this we talk about the rod-to-stroke ratio. This is the center-to-center length of the connecting rod divided by the stroke of the engine. There is quite a variance in stock small-block Chev-

rolets, which poses the question of what difference, if any, these various rod-to-stroke ratios make.

To begin with, we must establish an accepted working value. Rod-to-stroke ratios much below the 1.7:1 mark are considered by most authorities to be less than desirable in terms of power output. So why, with all the knowledge on hand, do manufacturers such as Chevrolet produce engines with shorter rod-to-stroke ratios? The answer is packaging, and the fact that the engineers who originally designed the engine never, in their wildest dreams, realized that the small-block Chevrolet would be so successful for so long.

Remember, when the small-block Chevy was first introduced it was only 265ci and had a rod-to-stroke ratio of 1.9:1. For a production engine this is more than acceptable by any standard. Although a greater ratio may have been even more acceptable, a law of diminishing returns applies here, just as in many other areas of engine design. For a production en-

gine, taller blocks to accommodate longer rods mean a heavier engine, more material and machine work and so on, thus the 1.9:1 rod ratio for a 3.0in stroke looked to be a sound choice.

As time went on it became apparent that Chevrolet really hit on something with their small-block, and its popularity intensified along with the demand for more horsepower. The easiest way to obtain more streetable horsepower is through more cubic inches. Ultimately, the little 265ci engine grew to 400ci, and all the while the deck height remained constant at 9.025in.

A 400ci engine sports a 3.75in stroke and to get that much stroke into the engine some compromises were made in the rod department. This entailed a reduction of the connecting rod center-to-center distance. With a rod only 5.56in long, the 400 engine had a rod-to-stroke ratio of only 1.48:1. Why the factory went to a shorter rod was not clear because a 5.70in rod can still just be fitted into the confines of a small-block Chevy, even with a 3.75in stroke.

Regardless of why it was done, for little more than the price of a set of stock rods and the dyno time involved, we are able to test the difference between a 5.56in rod and a 5.70in rod in this engine. In terms of change, going from 5.56 to 5.70in rod length increases the ratio from 1.48:1 to 1.52:1. Not a great deal of difference, but nevertheless a step in the right direction. The test engine involved was a 383ci small-block Chevy.

From the mid 1980s onward the 383ci enjoyed a great deal of popularity. It was most often produced by grinding down the main journals of a 400 crankshaft to the size of those for a 350 for installation into a 350 block. This, in effect, produced a long-stroke 350. For the most part, people selected the 400 rod to use in this combination because it cleared the cam lobes, and several relatively

Rod-to-Stroke Ratio Comparison

CI	Stroke (in)	Rod-Stroke Ratio
262	3.100	1.84
265	3.000	1.90
267	3.480	1.64
283	3.000	1.90
302	3.000	1.90
305	3.480	1.64
307	3.250	1.75
327	3.250	1.75
350	3.480	1.64
400	3.750	1.48

Check these figures and you'll see that the original 3.000in stroke motors had a virtually ideal rod-to-stroke ratio of 1.9:1. In an effort to retain commonality, this almost ideal rod-to-stroke ratio went down the drain as the small-block got bigger.

Short Versus Long Rod Comparison

	5.56in Rod		5.70in Rod	
RPM	TQ	HP	TQ	HP
2500	341	162	341	162
3000	356	203	356	203
3500	359	239	363	242
4000	338	257	338	257
4500	309	265	310	266
4750	289	261	292	264
5000	253	241	257	245

Here we need to consider trends rather than absolute results, and this test shows a trend for the power to increase as rpm goes up. However, we did not see a gain in torque at low rpm, which means that the theoretical increase due to the short rod was used up in friction between the cylinder wall and piston. Changing rods also meant a changed piston configuration, which can also affect the final results.

80

inexpensive pistons were available to simply bolt the assembly together.

To install the 5.70in rod it was necessary to machine quite a bit off the top of the piston, leaving the crown a little on the thin side and the top ring land somewhat close to the top of the piston. However, it was deemed reliable enough to attempt a brief dyno test.

For the figures in the chart, the major change was the length of the connecting rod. Such things as compression ratio and so on were held constant. Since the differences were expected to be relatively small due to the small change in rod-to-stroke ratio involved, these tests were conducted very carefully. Apart from keeping all contributing factors fixed, other than the ratio being tested, the before and after figures are the average of five tests. As you can see, the trend is for the longer rodded motor to make more horsepower even though, in this instance, we're only considering the effect on a relatively mundane, low-output street motor.

Another factor came to light, however, which doesn't show on the chart but is relevant for a street motor. Though likely to be affected by other factors, it was found that the 5.70in rodded engine exhibited less mechanical noise than the 5.56in. This could be due to the small reduction in rod angularity of 19.7deg. for the 5.56in rod, as opposed to the 19.2deg. for the 5.70in rod.

Long Versus Short Rods

At this point we've established, with a reasonable degree of certainty, that a rod longer than that of most small-block Chevys is desirable. Let me explain why. Consider connecting rod length in terms of piston motion. The chart shows the piston motion away from TDC and BDC with two rod extremes possible in a 350ci small-block. Notice that the long rod decelerates toward and accelerates away from TDC more slowly than the short rod, whereas at BDC the situation is reversed. Your initial reaction to this may be less than overwhelming, but there are numerous favorable implications.

First, the long rod reaches its peak velocity later in the downward stroke than the short rod. Second, the peak piston speed it generates is less.

A real-world example will illustrate the differences. Let's take a 6.0in rod

in a 350ci Chevy turning 7000rpm. Peak piston speed occurs at 75deg. ATDC (after top dead center) and at this point the piston is traveling at 6,642fpm (feet per minute)—approximately 75.5mph. On the other hand, the 5.7in rodded motor reaches peak piston speed at 74deg. ATDC and peak piston speed is slightly higher at 6,670fpm. The demand made for airflow on the cylinder head through a valve which at this point is still opening (full open is usually 102-106 deg. ATDC) is dictated by the area of the piston and the piston velocity.

By and large, a small-block Chevy is an engine short on breathing, especially at larger displacements. The biggest demand on intake flow occurs at the highest piston velocity. Although it is possible for the port and cylinder to play a catch-up game, to an extent it is preferable to satisfy a cylinder's demand for air at the time the demand is made.

When an engine is restricted by inadequate valving, the longer rod

does two things: it cuts the piston speed, thus reducing the instantaneous demand on the valve to flow air, and it delays peak piston speed

Rod Power Comparison		
RPM	HP 5.700 in Rod	HP 6.000 in Rod
5000	428	430
5500	461	462
6000	501	506
6500	522	527
7000	534	539
7500	512	518

Having an adequate rod length in a small-block Chevy makes for increased power. The changes may be small, but the longer rod has the ability to produce extra power and at the same time cut bore and ring wear. These figures are the percentage average gain from three different engines applied to one base line engine, so must be viewed as a trend.

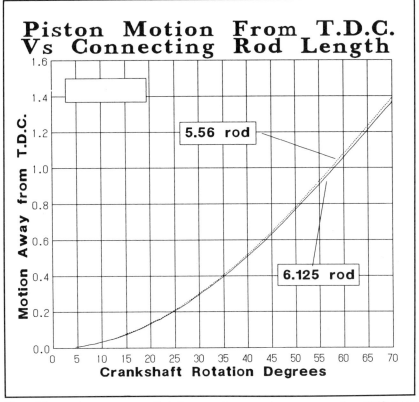

Piston Motion From T.D.C. Vs Connecting Rod Length

5.56 rod

6.125 rod

From this graph you can see that there isn't a lot of difference between a 5.560in rod and a 6.125in rod when their distance from TDC for a given crank rotation is plotted. Though the difference may look small, it is most certainly significant. The
longer rod moves away from TDC more slowly than the short rod, and this gives the intake valve more time to reach a higher lift value before maximum piston speed is reached.

until a little later in the cycle. This gives the valve more time to open and deliver the goods. The differences are small and it's difficult to say how significant they may be because the long rod has other assets that cannot be isolated during testing.

There is also a general trend for the volumetric breathing efficiency of the engine to improve slightly when a longer rod is used. This alone can be used as an indication that *slightly* more power is being developed in the cylinder.

Apart from its effect on the dynamics of the gas flow in the engine, the longer rod also exhibits less peak accelerations. Again, as an example let's take our 5.70 and 6.00in rods in a 350ci turning at 7000rpm. Peak piston accelerations with the 5.70in rod amount to 101,702fps (feet per second), whereas the 6.00in rod produces only 100,512fps. Admittedly it's only a 1.2 percent difference, but it's a move in the right direction. It will also help reduce crankshaft loading if the overall reciprocating weight produced by a longer rod in conjunction with a shorter deck height piston doesn't exceed its short rod counterpart.

Rod Angularity

The next area for consideration when making a long versus short rod comparison is rod angularity. Again, using our 350ci engine with a 5.70 and 6.00in rod as an example, we see the short rod has greater rod angularity. With a 3.48in stroke, the 5.70in rod produces 17.77deg. for a maximum angle with the 6.00in rod producing only 16.86deg. Though only about 1deg. less, this does represent a reduction. To appreciate why this is significant we need to look at the consequences of excessive rod angularity in an engine.

Generally, the greater the angle made between the cylinder bore and the crankshaft, the greater the side load is on the piston. A piston and ring assembly has one of the greatest friction levels of any part within the engine. The greater the rod angularity, the greater the amount of power sapped up in bore friction. In fact, excessive rod angularity due to too small a rod-to-stroke ratio proves to be one of the Achilles heels of big-inch small-blocks.

A good example here are some big off-road motors. At around 430ci, these engines are expected to go 500-1,000 miles in an off-road race. In spite of the fact that about 6000rpm is maximum with these motors, and air filtration systems used are the best money can buy, these engines often show severe bore wear and scuffing. Since pistons account for about three-fourths of the frictional losses in an engine, it's important to minimize it.

Higher piston speeds and rod angularity figures both contribute to intensify piston friction. Using a longer connecting rod reduces both these factors, hence more power is available at the flywheel because less is lost in friction. Just how much is gained in terms of flywheel horsepower by reduced piston friction is difficult to say, but some reasonable estimates would put this at 1.5 percent.

Another point in favor of the long rod concerns piston design. Although some piston designers favor putting the pin somewhere center to the skirt, leakdown tests have indicated that pistons with pins farther up than this tend to exhibit less leakdown. Although other design parameters may have an influence, the positioning of the pin as near the top of the piston as possible cuts the amount of piston rock. On race and high-performance engines this is significant because piston clearances are wider. Of course wide piston clearances are used to cut bore friction, but if it is at the expense of a gas-tight seal, then there is obviously a trade-off involved.

Anything that can be done to improve the seal by holding the ring more square to the bore should be done. A long rod with a pin higher in an appropriately designed piston helps to achieve this.

What's been discussed so far may seem like adequate justification for a longer rod in a small-block Chevy, but we haven't finished yet. Looking at the chart, we can see the difference in the rod angularity around half stroke of a 5.7in versus a 6.00in rod in a 350. Notice that the rod angularity with the longer rod is less. To get cylinder pressures to most effectively act on a crankshaft, the forces should be applied when the connecting rod is 90deg. to the crankshaft. The more the rod deviates from this idyllic situation, the less

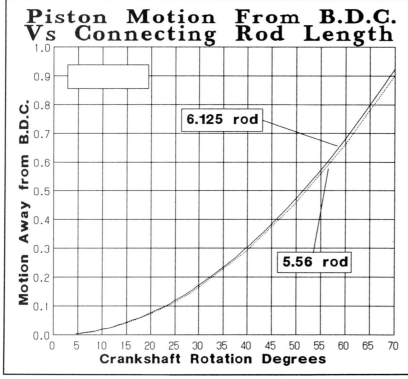

At the other end of the stroke, the long rod moves away from BDC quicker than the short rod. The short rod tends to hang around at BDC—that's a good asset to increase cylinder filling, but it's more than offset by the higher friction level given by a shorter rod.

effective it is at transmitting the forces.

By cutting rod angularity, longer connecting rods maintain a more favorable angle for a longer period of time between the crankshaft and the connecting rod. This is relevant when you consider that toward the end of the power stroke, the gases in the cylinder have dropped down in pressure appreciably from their maximum pressure. It is more difficult for a short rod to extract the energy from the expanding gases toward the end of the power stroke than it is for a long rod. The difference is small, but it is nonetheless there.

At the other end of the power stroke we see a different situation. Because the crank turns a greater number of degrees for a given movement of piston at the top end of the stroke, ignition timing should be set to take place a little later. The high cylinder pressures generated at the beginning of the power stroke also have a bearing on the piston and are transmitted to the crank for a little longer in terms of crankshaft rotation. The bottom line, then, is that a longer rod tends to extract more power from the expanding gases than its short counterpart.

Are there any redeeming features of a short rod? It's possible to answer yes, but any advantages to a short rod are limited and completely outweighed by disadvantages. The chief advantage is that with a short rod the piston hangs around longer at BDC. This means any negative effects of delayed intake closure are reduced. If we look at things in terms of piston movement the intake valve can be closed later in terms of crankshaft degrees for only the same amount of piston movement up the bore. As a result, the engine can be cammed slightly longer to take into account top-end performance without damaging low-end performance.

But how much is this worth? The difference in a typical point of closure from a 5.56in rod to a 6.00in rod is only a little over a degree so this asset of the short rod is minimal and to a large extent can be discounted.

Advantages of a Long Rod

So much for theory. Now let's see what gains are possible from a long rod. An interesting analysis concerning rod length was instigated by Peter Saueracker, technical editor of *Circle Track* magazine, with the feature running in the March 1990 issue. Research for this article was developed on a Superflow 901 dyno at Air Flow Research. The engine concerned was a 3.00in stroke unit and utilized 5.50 and 6.50in rods for the test.

On an engine producing almost 500hp the long rods proved superior to the short rods to the tune of 15hp. Since air consumption made only a minor and almost insignificant change in the upward direction, but fuel consumption improved measurably, it's reasonably safe to assume that most of the gains were brought about by reduced internal friction.

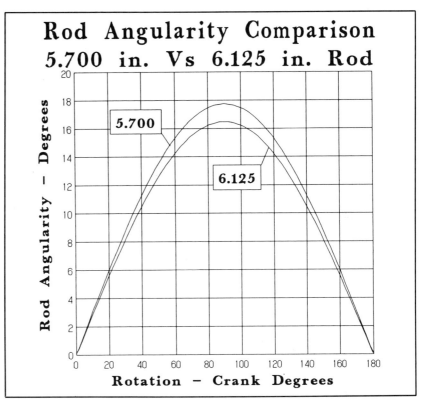

Though the effect on piston position with regard to crank angle is small you find when rod angularity is investigated the difference between a long and a short rod is much greater. A long rod produces less angularity.

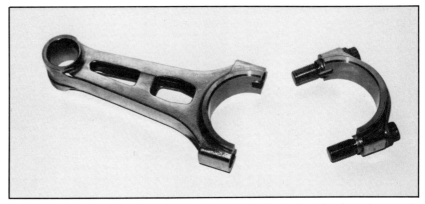

To date, the lightest steel rods that I've seen are these cast-stainless Mechart rods. This company produces several variations at different weights, and hence different strengths, for various applications. We used a set of 490g rods in a restricted two-barrel motor, they made for a low bobweight, resulting in small counterweights on the crank. The engine proved competitive until after about a dozen races, the car was totalled.

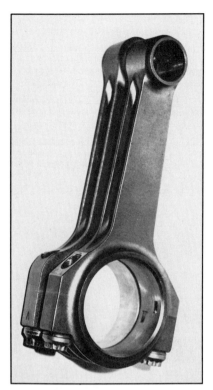

These Carrillo rods were used in a 406ci small-block. Although on the heavy side, they are undoubtedly a stout connecting rod. To clear the cam with a 3.750in stroke, it's necessary to grind on the shoulder of the rod, as seen by the upper rod of this pair.

Long Rods for Small-Blocks

Undoubtedly the long rod has proved itself, but what does this mean to a typical 350 small-block? Given a 3.50in stroke in the confines of a regular small-block it is difficult to increase the rod length more than about 0.400in. The most popular move when building a 350 small-block Chevy is to go from the stock 5.70 to a 6.00in center-to-center length rod. But how much is such a modification worth in terms of horsepower?

The graph shows the power curve for a 6.00in rodded motor, and the dotted line shows the curve of a projected 5.70in rodded engine. Though based on test results, we should look at this curve as being only a strong indication of the advantages of a 6.00in over a 5.70in rod. For what it's worth, the improvements shown by the longer connecting rod were substantially supported by the figures shown in the *Circle Track* feature.

A look at all the different types of motor sports where small-block Chevrolet engines are used will reveal that the 350 is not used exclusively. Due to race rules, many small-block Chevrolets are run at displacements other than this popular size. If space exists in the block, it's worth-

while posing the question of how long a rod ratio is optimal. Obviously the law of diminishing returns applies here, which leads us to ask where we should stop worrying about accommodating a longer rod in the engine. Anything below a 1.9:1 rod-to-stroke ratio should be considered less than satisfactory, but measurable gains still appear to be possible with ratios greater than 1.9:1. This is especially true on high rpm engines where ratios up to 2.2:1 are proving beneficial.

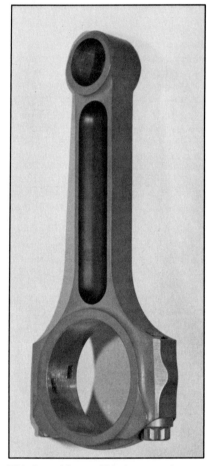

This fancy blue and black connecting rod is a Callies rod that has been finished with Swain Industries polymer oil-shedding coating. (The blue and black color scheme was simply for the benefit of the front cover photo.) Should you want coated rods, it'll have to be either blue or black, as Dan Swain's only into rod decorating for special occasions! I'm impressed with the Callies rod not only because of it's capability of running 7 million plus cycles, but also because this 6.125in version weighed in at only 640g, making it one of the lightest endurance rods on the market.

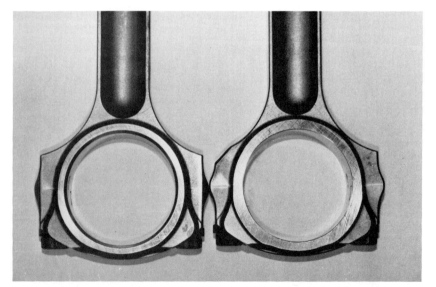

Note the ears by the bolt head on the cap flange. Some rod manufacturers such as Crower use ears on the rod side of the flange also, to compensate for the bolt hole. However, the presence of the bolt hole doesn't cut down the amount of metal supporting this area quite as much

as the bolt head. As a result, the top ears are not as critical as the bottom ones. These Callies rods went over 7 million cycles on a rod-test machine. At the load level used, anything over 4 million cycles has proven to be more than adequate for Grand National Winston Cup engines.

Aftermarket Steel Rods

There are numerous designs of good aftermarket rods available. Many are intended to fill a particular niche in the high-performance field and it pays to know what to look for regarding your application. At this point we will deal with steel rods only because aluminum rods are a whole new ballgame and need to be dealt with separately.

To begin with, do not overlook the fact that rods can be expensive. A look at the connecting rods available for a small-block Chevy reveals a basic trend. As a rule, the stronger the rods are and the heavier they are, the more expensive they are. Sure, there are a few rods that are the exception. But if you assume that this basic rule applies most of the time, then rod selection will be more cost effective.

The plan, in effect, is to not put any more rod into the engine than is needed—that is, unless an unlimited budget is available. There is no point in putting in a maxed-out rod capable of sustaining over 8000rpm at 650hp

levels indefinitely if your engine's only going to make 500hp and it's only going to be used for drag racing once in a while.

The key to success is not *over-engineering* the bottom end with your connecting rods because overengineering normally means *overweight*. To be able to run the lightest, and usually the cheapest, rods it's necessary to choose the lightest piston that the engine can get away with and yet remain reliable.

The point is to buy the rod to suit the rpm and power band that the engine will run in. It's a good idea to look at all the connecting rods available. Get the weight and material specification from the manufacturer, along with its price, and then make your decision.

I've used a substantial number of Carrillo rods without breakage, but have always felt that they're a very heavy piece.

I've had some limited experience with the Manley Carrillo look-alike on some high-output engines, and over a period of about three years, one set of Manley rods was used in several engines and performed about

the same number of revolutions as one would expect from about 1½ Daytona 500 races. The rod dealt with power levels ranging from 450 to about 900hp.

In nitrous-oxide motors up to 1,000hp, I've used rods from RHS. Again, no failures.

Lately, I've taken a hard look at the Callies connecting rod, as independent fatigue tests have shown them comparable to the Carillo rod, but considerably lighter.

The last rod I would like to mention is the Cosworth rod. Used by several successful Winston Cup engine builders, this rod looks a lot like

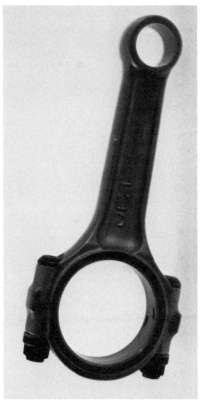

Here is an orchestration of a connecting rod in titanium. Both Jet and Crower make titanium rods. The rod here, together with the rest of the set, went into a Pro Stock motor for test purposes. The car was consistently launched at 10,000rpm plus, and after about four years I simply lost count of the number of passes it made down the strip. I lost contact with the test vehicle in the late 1980s and to date I've not heard if any of these rods have broken. Jet will custom manufacture these rods to order, but expect to pay a little over $400 per rod. Depending upon the length and application, I've seen these titanium rods down to 405g—and that's light!

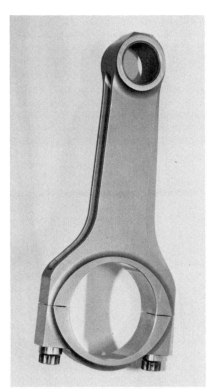

Here's Cosworth's offering for the small-block Chevy in the way of connecting rods. This is being used by some successful Winston Cup teams, who like the fact that apart from it's reliability, it also weighs in around 40-50g lighter than a typical Carrillo.

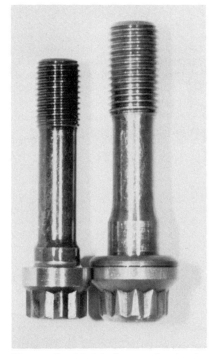

The small rod bolt is the type fitted to the Cosworth rods. The larger one is typical of that fitted to most other rods. The weight saving here amounts to about 8g per bolt, that's 16g right off the big-end weight.

a Carrillo, but has been through a weight reduction process. Unlike most of the connecting rods, it employs a $^3/_8$in bolt rather than a $^7/_{16}$in.

One of the lighter aluminum rods is the Super Rod. I've had only limited experience with them, but have never experienced a failure. This brand of rod has been used in some of the most successful drag-racing small-block Chevys, but does need to be changed regularly.

By using the finest aerospace-quality bolt, Cosworth has been able to lighten the rod a bit, especially around the big end, which means less counterweight. A reduction of just 5g from the big end results in a 10g drop for a counterweight. Any change in the big-end weight is doubled in terms of a reduced bob weight to balance the crank.

Aftermarket Aluminum Rods

As a connecting rod material, aluminum has both advantages and disadvantages. Not only is it light, but its strength-to-weight ratio is good, another primary consideration for a connecting rod. However, aluminum fatigues easier than steel and its expansion rate is higher. As a result, aluminum connecting rods need to be initially overdesigned to produce a reasonable fatigue life.

The heat treatment on an aluminum rod is also important because this can ultimately have a tremendous effect on its life. Most connecting rods go through a T6 heat treatment. As far as thermal considerations are concerned, aluminum rods can grow 0.006-0.007in more than a steel rod, so piston-crown-to-cylinder-head deck clearance needs to be adjusted suitably when measured cold to compensate for thermal growth.

Under high-rpm conditions the amount of stretch an aluminum rod experiences is usually more than a steel rod, but since the cross section of aluminum rods varies greatly between one manufacturer and another, the amount of stretch occurring is also a variable. Consult with the rod manufacturer for the minimum pis-

ton-to-deck clearance you can get away with. Give them an idea of the stroke, the weight of the piston and the rpm the engine is expected to turn.

Aluminum rods are generally used in drag-race engines because of their light weight. The tendency has been to replace the rods about every 75-100 runs down the strip. One aspect needs to be considered, however. Drag-race engines are usually maximum-effort engines, and the rpm is far higher than on any other type of usage that the small-block Chevy may be exposed to.

Although aluminum rods have a limited fatigue life, one must consider the amount of rpm they're subjected to. It doesn't take much of a drop in rpm to considerably extend the fatigue life of an aluminum rod. If a beefy enough rod is chosen in the first place, then there's every possibility that rod life may be extended so much that it can justifiably be used for applications other than drag racing.

For instance, we know that the Bill Miller rod has been successfully used

Because of the greater rate of expansion of aluminum rods, bearing retention is more critical. Aluminum rods typically *have a pin, such as seen here, to reduce the probability of a spun bearing.*

Rod Bolts—How Big Is Big Enough?

For whatever reason, most racers work on the bigger-is-better system and when looking to buy connecting rods, view the connecting rod bolt in much the same light. The truth of the matter is, most aftermarket rods are overbolted.

When designing their Chevrolet rod, Cosworth showed that for most applications, a $^5/_{16}$in bolt of a suitable high-grade material was more than adequate for the job. The difference in cross-sectional area between a $^5/_{16}$ and a $^7/_{16}$in bolt, which is the common size, is some 90 percent. If a $^5/_{16}$in bolt will do, why do most rod manufacturers see the need to almost double the size? The answer to this is part politics, part mechanics and part economics.

To make a $^5/_{16}$in bolt work it must be made of the best materials, and any margin of error must be diminished. A bigger bolt made of a slightly lesser grade of material proves to be stronger, less expensive and more break resistant. Because the racer perceives bigger is better, it is easier to sell a rod with a big bolt. Even Cosworth conceded this by putting their rods into production with a $^3/_8$in bolt instead of the original $^5/_{16}$in.

in sprint-car engines with, of course, some regular changes to eliminate the possibility of breakage. But moving down the scale a little, some of the classes that have restrictive breathing can utilize a rod of this type for a typical season's racing. The advantage is that the rod is much stronger than a stock steel one, and usually around 200-220g lighter. By the time you've paid all that money out to doctor the stock rods, it doesn't take much more cash to buy a good set of aluminum rods.

As far as street motors are concerned, a good stout set of aluminum rods should not be dismissed as totally unsuitable. If the rpm limits of the engine are 7200-7500, then in a typical 350, which is driven normally most of the time and maybe makes some full-throttle runs equal to say ten passes down the strip a week, the rods could last almost indefinitely—certainly 15,000 miles would not be unreasonable.

Although I've not used aluminum rods in Chevrolets, I've had some limited experience in putting aluminum rods into Pontiac V-8 engines. In stock form the Pontiac rod is cast iron and is subject to a much higher than average failure rate. At the time of writing, a set of Bill Miller rods has been in a street-driven Pontiac for three years. In the Pontiac application, at least, it would seem that hundreds of engine builders have successfully used an aluminum rod for street use.

The key, as ever, is making sure the rod is adequately strong. This means choosing a rod that has adequate cross-sectional area in the places that matter. If the rod is sufficiently bulky in the beam, then breakages usually occur from the shoulder into the big-end bore or from some point at the thread into the big-end bore. Alternatively, a cap failure may take place and this would normally be through the edge of the counterbore, from the bolt hole to the big-end bore.

When building a high-performance small-block Chevy, I like to set the side clearance on steel rods at around 0.015in or so. When an aluminum rod is used, this clearance is stepped up by 0.002–0.003in to compensate for the normally greater expansion seen at this point with an aluminum rod. As far as small-end clearance goes, I normally use 0.0004-0.0006in, which opens up considerably when the rod reaches operating temperature.

At the big end of an aluminum rod certain features become more critical than on a steel rod. Because of the greater diameter of the big end, and the fact that the rod expands more, retention of the bearing in an aluminum rod is more difficult. Because of this, it's common practice to put a brass retention pin into the rod cap to reduce the possibility of the rod bearing spinning. However, the appropriate bearing crush needs to be maintained with the brass pin as well.

A permanent ovalizing set in the big end may, with some brands of aluminum rods, be an early sign that the rod is nearing the end of its life. Whenever the engine is stripped, it's a good idea to check the size of the big-end bore, both vertically and horizontally. The housing itself should be round, but don't be misled into thinking that the bearing ID should necessarily have a truly circular bore. Most bearings are cut away at the parting line so as to give a larger clearance figure. Typically, between 0.006–0.008in clearance will exist at the parting line when 0.002–0.003in are shown vertically. This disparity in diameter between the vertical and horizontal is deliberate to compensate for ovalizing of the big-end housing under high-rpm applications for *any* type of rod. Since the stiffness of variously designed aluminum rod housings differs quite markedly, it's a good idea to check with the rod manufacturer for the acceptable limit if any obvious ovalizing exists.

When assembling aluminum connecting rods be sure to use the thread lubricants recommended by the rod manufacturer. Another important factor is the precision of your torque wrench. Overtorquing the bolts in an aluminum rod can cause a thread failure in the rod; undertorquing them by any significant degree can cause a bolt or housing failure. If an aluminum rod has been through an engine blowup, watch for nicks in critical areas caused by flying debris. If any occur, they can be a prime source of future cracking, though small imperfections can often be polished out.

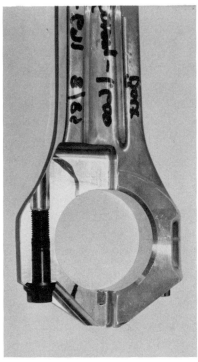

The critical area for most connecting rods is as indicated here. If they're going to fail, it's usually from the flank radius of the beam into the big end.

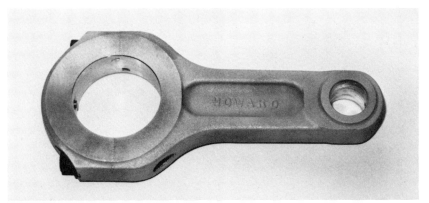

Howards have been making aluminum rods for a long time, and over the years I've used two or three sets of their rods in motors, all bracket-race applications. With an 8,000rpm limit on one engine, I've seen over 200 passes without failure.

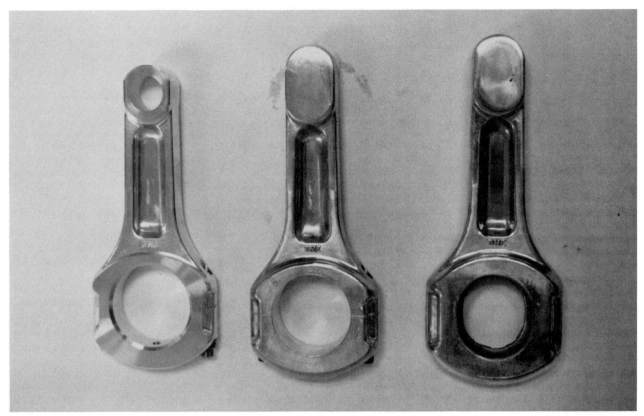

Here's a typical process from the forging blank to the finished machine job on an aluminum connecting rod. Sure, there may be more than three operations to get from one to the other, but it does give you an idea of how things progress.

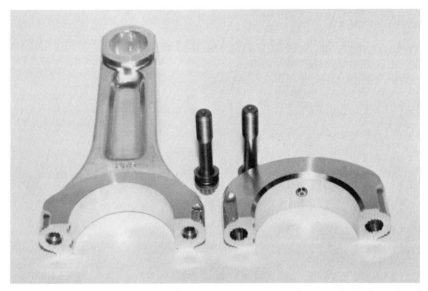

Bolt size can be more critical on an aluminum rod than on a steel rod; a large bolt is needed because a smaller one would simply pull the thread from the aluminum rod. As a result, the integrity of the big end is much more dependent on the rod bolt size and the strength of the thread in the rod.

Chevrolet Cast and Forged Crankshafts

9

Because of availability and price, most hopped-up small-block Chevrolets are going to utilize a stock or near-stock factory crankshaft. Therefore, it's a good idea to know what's out there. Fortunately there's good, there's mediocre, but there is nothing really bad. Generally speaking, factory cranks are good news, even if you are trying to extract relatively serious horsepower from your engine.

The question that needs to be addressed up front is that of crank-to-block interchange. Since the small-block Chevy was introduced in 1955, there have been a number of changes made in the bottom end. Some work in our favor—but others can prevent easy interchange. Obtaining this information is important before embarking on anything other than a straight engine rebuild.

Cast Versus Forged Crankshafts

Probably the most significant difference in crankshafts is whether the cranks are made from forged steel or cast from nodular iron. For the inexperienced engine builder, the first thing to learn is how to tell the difference between one and the other.

When forgings are made there is a large amount of what is called flash along the forging die split line. The flash is sheared off in a secondary stamping operation. This usually leaves a mark $1/4$–$1/2$in wide along the split line of the crank dies. Because it greatly simplifies the design, the forging die has a horizontal split line, so it's necessary to forge the rod journals on a single plane. After being removed from the die and deflashed, but while still at forging temperature, the cranks are twisted at numbers two and four main bearings to put the rod journals in the right angular position.

In casting a crank, the metal can be put where it's needed in a far more precise manner. The split line left by the casting mold tends to be much finer than on a forged crank. As a

result, we normally see only a split line of some $1/16$in wide.

All cranks made before about 1963 were forged, using SAE 1046 alloy steel. After being forged, the cranks were heat treated. Supposedly, the intention was to give a little additional hardness to the surface and refine the grain structure of the core of the crank for added fatigue resistance.

In 1967, a change in material specification was introduced, which coincided with the introduction of the 350. Instead of using the SAE 1046 material, the cranks were manufactured from SAE 1053. Unless Tuftrided, no subsequent heat treatment was done on these cranks after forging, other than any stress relieving that may have been required.

The factory made cranks for some high-performance applications that were given a Tuftriding process. This process produces a little extra in the way of surface hardness, but its greatest asset is that it increases the fatigue resistance of the cranks substantially—figures of as much as 150 percent are bandied about the industry for its effectiveness. An additional

advantage is that a Tuftrided surface is scuff and wear resistant.

Cast cranks were introduced about 1963, which more or less coincided with the introduction of the 327. As Ford Motor Company had demon-

Here is a cast crank. It can be identified as such by the thin parting or flash line on the casting pattern joint.

CI	Year	Main Bearings (in)	Rod Bearings (in)	Stroke (in)
262	1975-76	2.450	2.100	3.100
267	1979-80	2.450	2.100	3.480
283	1957-67	2.300	2.000	3.000
302	1967	2.300	2.000	3.000
302	1968-69	2.450	2.100	3.000
305	1976-on	2.450	2.100	3.480
307	1968-73	2.450	2.100	3.250
327	1962-67	2.300	2.000	3.250
327	1968-69	2.450	2.100	3.250
350	1967-on	2.450	2.100	3.480
400	1970-72	2.650	2.100	3.750
400	1973-80	2.650	2.100	3.750

Chevrolet Crankshaft Specifications

Here are the basic crankshaft specifications used since the small-block Chevy came out. Big motors are better off with big-journal cranks simply because they need the strength. Smaller-displacement motors can usually make more power on the smaller-journal cranks as they exhibit fewer bearing losses.

strated for many years, cast cranks, if made with suitable materials and techniques, were long-life pieces even when relatively high power outputs were involved. The type of material commonly used for cranks is nodular cast iron. This differs from regular cast iron in the way the graphite crystals lie in the grain boundaries of the parent metal.

The flash line on a forged crank is relatively wide; what you see here is typical. This pattern is left when the shearing dies remove the excess metal left along the forging split line.

It will probably help to first understand what cast iron is. Essentially, cast iron is iron that has an excess of carbon in it. If you take plain iron, having no carbon in it, and add a little carbon, then it becomes a steel. A high-carbon steel is a steel having a little less than 0.9 percent of carbon in it. Carbon in excess of this amount will not dissolve into the iron. Whatever additional carbon that exists, forms at the grain boundaries and causes the steel to take on the properties of cast iron, becoming less ductile and more brittle.

On the positive side, the additional carbon allows the iron to be cast easier. Most cast irons have 2-3 percent carbon in them. In its normal form the excess carbon in cast iron appears at the grain boundaries in the shape of an irregular flake or crystal. By alloying the cast iron with certain elements it is possible for the excess carbon to form a spheroidal or nodular particle of carbon at the grain boundaries. Carbons that take a spherical form rather than a randomly irregular form produce a cast iron having greater tensile strength and fatigue resistance.

The excess carbon in the iron also acts as an internal lubricant. This is why cast iron is used for lathe and milling machine bed ways and the like; the excess carbon, or to use the correct term, graphite, dramatically improves the wear properties in comparison with raw iron or steel.

A regular type of cast iron is highly unsuitable for cranks. The irons used for crankshafts are alloyed with various other materials which toughen and strengthen the base material. These materials are typically manganese, nickel, chromium and vanadium. The bottom line for the high-performance engine builder on a budget is that Chevrolet cast cranks are dependable, good pieces. Indeed, in some respects, they can hold their own with a forged crank. In fact, it's worth making some comparisons here to put things into perspective.

In terms of outright strength, the forged crank is stronger than the cast crank; however, the free graphite in the cast iron makes it more wear resistant than a typical forged crank, unless the forged crank has been specifically heat treated to form a wear-resistant surface. On the other hand, if the cast crank is also given suitable heat treatment to enhance wear resistance, it can bounce right back on a par with a forged crank.

Another plus for a cast crank is that up to certain stress levels it is less prone to crack than a forged crank. This means that so long as reasonable rpm and horsepower limitations are observed, the cast crank is more than acceptable for use in a high-performance motor.

Journal Sizes

In 1955, the forged cranks in the 265ci small-block had journals sized at 2.00in for the big ends and 2.30in for the main bearings, used in conjunction with the 3.00in stroke. Until the introduction of the 327 in 1962, the 3.00in stroke was common. The 327 achieved this extra capacity by utilizing the 4.00in bore, up 0.125in from the 283, having a 0.250in longer stroke. Note that 327 cranks won't go into earlier blocks because there isn't enough counterweight clearance inside the block.

The introduction of the 350 in 1967 brought about a change in both journal and stroke sizes of the small-block Chevy crank. The stroke was increased to 3.48in and to give the crank what the factory thought would be the required additional strength, the main bearings and journals were increased in size to 2.45in and 2.10in respectively. When these larger journals were introduced on the 350 crank, at about the same time, other cranks were brought in line by having correspondingly sized journals. Es-

When the crank's installed and the mains all torqued to the correct value, you should be able to rotate the crank from the damper snout with just pressure between thumb and forefinger.

sentially, this means that all cranks produced after late 1967 or early 1968 had the big journals.

When the 400ci small-block was introduced in 1970, it sported a 3.75in stroke. As a result, the overlap of the big-end and main bearings, as viewed from the end of the crankshaft, was reduced because of the increased stroke. To compensate for this, on the 400 crank only the main-bearing size was increased from 2.45 to 2.65in.

Counterweights

Anytime an increase in stroke length is made at the factory, it's a safe bet that the thinking behind it was to get extra power and torque from the engine. Thus, there are many variations on the theme to suit various high-performance vehicles. For instance, when intended for higher output engines, the thickness of the arm between the main bearing and a big-end journal increased. The increase is most noticeable on the number one and number four arms connecting the front and rear big-end journals to the main journals.

As stroke lengths were increased, it was necessary to add more counterweight to the crankshaft so that the engine could be completely balanced. However, there's a limited amount of room in a small-block Chevy crankcase. Also, the retention of the near-universal 5.70in rod length and the limited deck height of the block meant that the underside of the piston made a close approach to the counterweight, even on a 3.00in stroke motor. Longer strokes used up that clearance and then some.

To generate clearance on longer stroke engines, it was necessary to cam-turn the counterweights. In other words, the center of the OD of the counterweight was not the same as the center of the main bearing and was, in fact, offset toward the big-end journal. This meant the distance from the middle of the counterweight to the center of the main bearing was less than from either the leading or trailing edge of the counterweight to the center of the main bearing. With the exception of the later big-journal 302 crank, all the 3.00in stroke cranks had circular counterweights, but the 327, 350 and 400 cranks used cam-turned counterweights.

In spite of adding as much internal counterweight to the engine as possible, more counterweight material was still required for the longer stroke engines to balance out the loads. Thus, a small amount of counterweighting was added to the flywheel flange. Early 3.00in stroke cranks had a round flywheel flange. Later 3.00in stroke cranks, such as used in the small- and big-journal 302 engines, had a round flywheel flange with a notch cut in it.

The 327 cranks can have either a small or relatively large amount of counterweighting on the flywheel flange. This additional amount of counterweighting is positioned on the same side as the rear counterweight.

On 350 and 400 cranks the amount of counterweighting on the flange is evident. Even so, with the 400 crank, the amount of counterweighting that can be accommodated is still not enough to fully balance the engine. For a 400 crank a special damper and flywheel is required, as these have additional counterweighting to bring the crank into balance. So unless it's physically changed, the 400 crank is what is known as externally balanced. Therefore, on a 400 crank you can only use a flywheel and damper made specifically for a 400ci engine.

In 1986, a number of changes were made to the small-block Chevy motor. Design changes relevant to the crankshaft included going to a one-piece real main seal instead of the split seal that had been used for so long. The one-piece rear main seal necessitated the redesign of the rear of the crankshaft. This type of crank had to be used with the relevant block, as the rear main seal was contained in a special housing that bolted to the back of the block.

3.00in Stroke Crankshafts

If you're going to use a stock crank for any application, then obviously the stroke length required is going to be the major deciding factor. So we'll run through the cranks in order of stroke length to give you an idea of which are best.

If you intend building a 3.00in stroke motor, then virtually any of the forged cranks will suit your purposes. Don't forget that forged cranks are available in both the small- and big-journal sizes, but because we're dealing with a short-stroke motor, the small pin diameters are going to be more than strong enough for almost any application, regardless of the crank you choose.

There is one provision though—be sure to have the crank crack tested. Don't forget, any of these early cranks could now have amassed considerable mileage and there's a good possibility that they're cracked—which may mean sorting through several cranks to find one that's usable.

Cast 3.00in cranks were available from about 1963 to 1968. Again, because of the short stroke, these cranks are usable to high power levels, but they are only available in small-journal sizes.

If you want to go with what is probably the best 3.00in crank, then you need to get hold of the crank used in the 302ci small-block between 1968 and 1969. These cranks were the big-journal type. Although the crank itself may not offer much advantage in terms of required strength over the small-journal one, there is an advantage in going to this size if you're sticking with a stock rod. The later big-journal rods are stronger than the earlier small-journal rods, so the combination of big-journal crank and stock rods may prove to be less expensive than a small-journal crank and an aftermarket rod. The choice is yours—it's one worth thinking about. Incidentally, many of the 3.00in large-journal cranks were Tuftrided, so this is an extra bonus if you decide to go this route.

If you're having a problem finding a small-journal forged crank that's any good and you decide to use a cast crank, then it's worth noting that most of the cast cranks were used in 1966-1967 passenger cars, while the forged cranks were used mostly in trucks. Don't try looking for a 3.00in big-journal cast crank because you won't find any.

3.25in Stroke Crankshafts

If a 3.25in stroke motor is your intention, then some interesting possibilities exist. Assuming you are looking for some high horsepower numbers, you'll need to make a decision between a small- or big-journal crank. First of all, there are no cast small-journal cranks, so you need not concern yourself with finding one. However, the small-journal forged cranks, if you select the right one, are good for drag-racing applications up to some pretty horrendous horsepower.

Grumpy Jenkins used reworked stock 327 forged cranks in his 700hp Pro-Stock motors. His preference was for the small-journal cranks because they had less frictional bearing loss. As he stated in his book, *The Chevrolet Racing Engine* (SA Design), he uses the 1962-1967 small-journal Tuftrided crank, part number 3838495. Since it comes from the factory, this is one of the heaviest forgings that Chevrolet made.

By contrast, the 1968 big-journal 327 Tuftrided crank is among the lightest made. If the intended output of the engine is not likely to overstress a cast crank, then one of the big-journal-only cast cranks could be used. This could be sourced from a 1968 and later 307 or 327ci motor.

3.50in Stroke Crankshafts

For most people the 3.48in stroke 350 crank would be the one that's most commonly used. These are available only in big-journal cranks in both cast and forged form. If you can get a forged one that shows up with no cracks, then great, use it.

The best forged crank to look for is the Tuftrided one, part number 3941184. Of course, that number's only good if you're going to buy a new one over-the-counter at your local Chevrolet dealer. If you want to look for one in a wrecking yard, then the most likely place to find one is in an LT-1 Corvette engine of 1969-1970.

These were the engines rated at 370hp and equipped with a mechanical cam.

If you cannot find a Tuftrided forged crank, it's no big deal. What you can do is find a good forged crank, although this is becoming increasingly difficult because most of them now have high mileages, and then heat treat that particular crank. Remember, before you do anything with any crank—forged cranks, especially—it must be crack tested.

Of course, other engines share the same stroke as the 350, though they may not displace the same. For instance, the 305 and the 267ci engines also have the 3.48in stroke. Dimensionally a 305 crank is the same as a 350, and these cranks are common. The 305 cranks are balanced to a lighter bob weight than 350 cranks, though, because of the lighter piston assembly on a 305ci engine. Typically, a 350 crank uses a bob weight around 1,925-1,950g, and the 305s can run approximately 25g less.

If you're using a slightly lighter piston than the stock 350 item, then a 305 crank will balance with no problem. Of course, if you're using a 305 crank it will be a cast crank, as forged 305 cranks are nonexistent.

3.75in Stroke Crankshafts

As far as 400 cranks are concerned, there used to be an abundance of them because 400 blocks wore out faster than the cranks. It became popular to use the surplus 400 cranks in 350 blocks. To do this it was necessary to grind the bigger main journals down on the 400 crank so that it would install into the main-bearings housings of the 350 block. With a 0.030in overbore on a 350 block, the 3.75in stroke of the 400 crank yielded 383ci.

If you intend using the 400 crank in a 400 block, then the good news is, for the most part, these cranks could not be overrevved in a stock motor, so chances are if they received regular oil changes, they would be in good condition.

There are no 400 forged cranks.

If you're going to the wrecking yard to get a 400 crank to install in a vehicle that currently has a crankshaft other than a 400, remember that the 400 crank is externally balanced. You will also need the appropriate damper and suitably counterweighted flywheel. If you've already paid for an expensive flex plate or flywheel, you do have another alternative if you're swapping to the 400 crank, and that's to use the ABS flex plate weight adapter. This fits between the flywheel and flex plate, and adds the necessary out-of-balance force to utilize a regular balanced flywheel for a crankshaft requiring external balancing.

Aftermarket Crankshafts

10

Anytime a stock factory crankshaft won't fulfill your engine's requirements, you're going to have to look at an aftermarket crank. Though there may be many other reasons for selecting an aftermarket crank, the two most common are increased durability and a stroke length normally not offered by a factory crank.

Crankshaft Materials

In an effort to put out stronger cranks, some years ago the factory introduced forging made from 5140 steel. In heat-treated form this material offers about 115,000psi tensile strength. Such forgings were available either as forgings or as finished cranks through Chevrolet parts dealers, but now only forgings are available for custom crank manufacturers to machine in whatever way they see fit. These forgings can accommodate stroke lengths from 3.00 to 3.75in.

Whereas 5140 usually proves to be adequate for most high-output applications, it's not considered strong enough to stand the rigors of a 650hp 8000rpm Winston Cup-style engine. The ability to run at such power levels and rpm for close to 500 miles without breakage has encouraged crank manufacturers to take a hard look at ways and means of building a better crankshaft than can be done in 5140. For this, most crank manufacturers turned to 4340 steel, which, in its heat-treated state as used in most cranks, is around 155,000psi tensile strength.

Trace Material		
	Comparison of 4340 and British EN40B	
	4340	**EN40B**
Chrome	0.70-0.90	2.9-3.5
Manganese	0.60-0.80	0.40-0.65
Nickel	1.65-2.00	0.40 maximum
Carbon	0.38-0.43	0.20-0.30
Molybdenum	0.20-0.30	0.40-0.70
Copper	0.18-0.20	—
Tensil mechanical properties tempered to 1100-1200deg. F. psi	118.7-138.8K	134.4-156.8K
Brinell Hardness	285-331	269-331

This chart shows the difference between 4340 and EN40B steels. At this time, Cosworth is the only company I know that uses EN40B for its crankshafts.

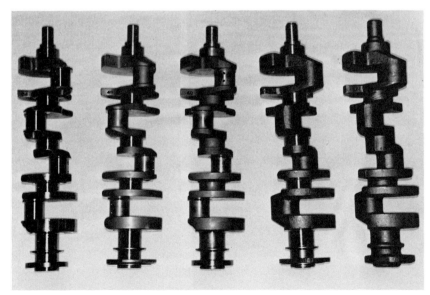

A crank goes through many stages from the original forging to the finished piece. Here is how the raw forging develops into the finished item as it goes through various machining operations.

The first machining operation on the crank is to turn the center main bearing to an accurately pre-determined size above the finished size.

A notable exception is the Cosworth crank, which is made of EN40B, a British steel developed almost specifically for crankshaft applications. After a nitriding heat-treatment process the EN40B cranks can, as with 4340, develop an extremely hard surface finish for good wear resistance. But apart from the finished hardness, there are indications that EN40B has a slight edge over 4340 in terms of strength.

Crankshaft Finishing

There's a lot more to finishing a crank than just a few simple machining operations. To find out what it

This accuracy is needed so that the crank can be held on this machined center bearing to turn all the other concentric diameters both in front and behind it.

With all the concentric diameters turned, the next operation is to load the semi-machine forgings into this pin mill.

takes, I went along to see John Callies of Callies Performance Products, Inc. This company is one of the larger manufacturers of aftermarket crankshafts in the United States.

Callies has built a reputation for itself by excelling in three areas: inspection, inspection and inspection! Every crank is numbered and goes through each process with an inspection sheet, and every single dimension on the crank, after it is machined, is listed on the sheet. Primary inspection is done by the operator at each machining station.

After a certain number of operations are completed, an inspector verifies these dimensions. When the crank is finished, it is inspected on an expensive AdCole automated checking machine which can define errors down to 0.000040in. It also produces a permanent record of the crank's dimensions, which are kept at the factory as a reference should there ever be a problem or a question concerning that crankshaft.

Starting from the raw forging, the first operation is to center drill the ends of the crankshaft to establish the centerline. From here, the crank goes into what is known as a Fay lathe, which machines the center main bearing using two tools—one coming in from each side to eliminate any flash in the shaft. Although this is essentially a roughing operation, it is done with as much precision as possible because initially, at least, all longitudinal dimensions are set from the faces of the center main.

Once the center main has been turned in the lathe, it is then rough ground. The rough grinding entails grinding this diameter to a precision size above the finished size. At this point that diameter is going to be used as a data and a drive, so it must be relatively accurate.

From here, the forging is set up in a special lathe that drives from the center main bearing; this machine roughs out all the remaining concentric diameters on the shaft. Once the main mass of metal has been removed, the forging is then transferred to a multi-wheel grinding machine, which rough grinds all the mains. From here it goes to an unusual machine known as a pin mill. The pin mill is probably unique to the crankshaft business. Although the crankshaft turns, the tools are mounted in what can be best described as a large iris and this spins

around the journal, closing down to machine the big-end journals to a rough finished size.

Next comes the drilling of the oil holes. A purpose-built machine takes care of this operation. When a hole is drilled directly across a shaft there's no real problem, but when holes are required to enter the journal at an angle, then the drilling becomes a delicate operation, one made even more difficult by the tough nature of the material normally used for crankshafts.

After the oil holes have been drilled, the crank moves to a CNC lathe that turns the counterweights to size. Hand-finishing is then applied to de-flash the journal arms and counterweights of forging lines.

At this point the crank has had a considerable amount of metal removed from it. Anytime you cut through the skin on a forging there's a possibility of the metal moving because the skin of the forging has a considerable amount of stress in it. Machining it away causes distortion. The way around this is to stress-relieve the crank so as to settle out all those stresses in the skin.

At Callies the crankshafts are stress-relieved by a mechanical process known in the trade as MetaLax. The parts are fed vibrations that are varied in frequency until the resonant frequency of the component is found. Then they are left to vibrate at this frequency, and this settles out any stresses in the part.

Other stress-relief techniques are available; many people use thermal stress relieving. With this technique, the part is simply heated to sufficient temperature at which the material becomes just plastic enough to settle out the stresses. Once a crank has been stress relieved, any minor amount of metal that may have to be removed is not likely to cause any further distortion from machining, so at this point finishing operations can begin.

The finishing operations start with finish-turning the pilot bearing, counterbore, flange, nose and so on. Then the crank goes on to a CNC machining center that machines the keyway, then drills and taps the flange bolt holes. The keyway has a small radius applied to the corners, since sharp-cornered keyways have been shown to produce stress raisers.

From here the crank goes to a machine that grinds the main journals, the gearstep, the crank nose and so on to finished size. Once the main bearings have been ground, the crankpins are ground.

Once all the grinding operations are completed, the crank is sent off to the polishing department. When Callies' cranks are polished, the operation is controlled with air gauges to make sure that the journals stay straight, and are not deformed by the polishing.

All the heavy-duty machining is now completed on the part, and it goes for heat treatment to harden the surface. Callies calls its particular process a plasma heat treat. Principally, it is a form of ion nitriding. The process—which utilizes relatively low temperature, nitrogen bearing gases at low pressure and high currents in magnetic fields—tends to strip the material of its outer layer of molecules, and then replaces them. Also in the process, the nitrogen infuses with the steel, forming iron nitrides which are extremely hard.

During the plasma heat-treat process there is a slight growth of the crank surface metals, but because it is restrained, it takes on a compressive load. This surface compressive load acts the same way that a shot-peened surface does, and goes a long way toward eliminating the start of surface cracks. The process not only gives the crankshaft extended bearing life due to a hard finish, but also extends the crank's fatigue life.

Although the plasma heat treatment in itself relieves a certain amount of stress to the crank, all Callies cranks go on to a further stress-relief process right after heat treatment. Getting rid of any resid-

ual core stresses in the crank produces a part with more fatigue resistance.

From here the crank goes into inspection to check for straightness because the heat treatment process can on occasion cause distortion. Part of the job of a good heat-treatment specialist is to be able to treat parts with a minimum amount of distortion. Special jigs are made up to position the cranks during treatment to eliminate distortion.

After the straightness check, the cranks are Magnafluxed to ensure no

cracks exist from any previous processes in the crank. Following the Magnaflux process, the crank journals go on to a micro-finish process, which puts a super-fine polish on the journals. This is the last machining operation before a final size inspection is done on an automated AdCole inspection machine.

Once the weight of the rod and piston set to be used with the crank is known, the crank goes through its balancing procedure. Here Callies offers two forms of crank balancing: regular balancing, as seen in most

This purpose-built machine does nothing other than drill crankshaft oiling holes, which at best is a delicate operation.

The pin mill is a special machine that roughs out the big ends.

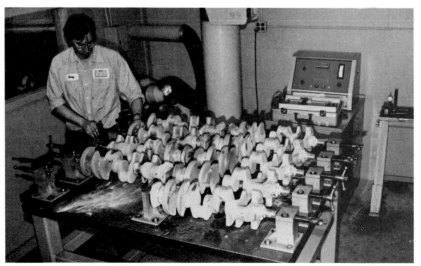

After all the rough machining is done, stress relieving of the crankshaft forgings is carried out on this Meta-Lax stress-relieving machine. It vibrates the parts at their natural resonant frequency to gradually settle out residual stresses.

balance shops, or, for an extra fee, a high-speed balancing.

For the high-speed balance, the crank is spun at the sort of rpm it is expected to see in-service—6000-8000rpm. By spinning the crank at this speed, it allows the crank-balancing technician to determine if material needs to come off any of the center counterbalance weights, as any out-of-balance force along the length of the shaft causes the crankshaft to deflect. At these high speeds, centrifugal loads can be developed to deflect the crank quite considerably, even if it's only out of balance by a small amount. The high-speed balancing ensures that the crank runs true in the block, even when spinning at 8000rpm. Normal crankshaft balancing jobs tend only to balance the crank at the ends, making the assumption that the crankshaft is infinitely stiff.

Twisted Versus Nontwisted Crankshafts

Over the years a certain amount of controversy has existed in the aftermarket crank industry as to whether a twisted or nontwisted crank is best. Factory cranks, when they're forged, have all the big-end journals on a single plane—this means they are not positioned 90deg. from each other, but 180. After being forged, the stock cranks are then put through another process that twists the crank at numbers two and four journal, so as to set the big ends (numbers one and two, and seven and eight cylinders) in the right relationship to the center ones.

For nontwisted forgings, the dies are more complex and possibly for engineering reasons that we're not aware of, seem to leave more metal in the forging than a twisted one. This ultimately means more time spent machining.

Mainly for cost reasons, the factory chose to use the twist method of producing the forgings, and the debate between which is better centers upon whether the twisting action causes a reduction in strength. My thinking is that since a forging is a process designed specifically to move

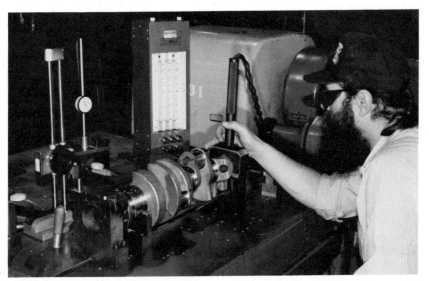

Inspection is an ongoing thing at almost every stage of manufacture. Here, journal accuracy is being checked at three points across the diameter with a special air gauge.

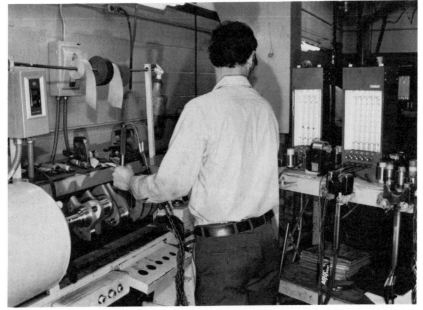

After final grinding, the journals are micropolished, and again this is done with due reference to air gauges to make sure that journal paralysm and diameter are held to tight tolerances.

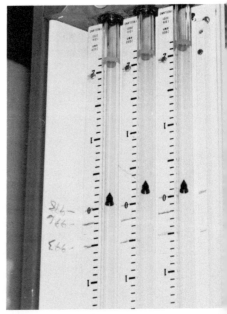

As you can see from the air gauge indicators, there's less than a tenth of a thousandths variation.

metal anyway, a simple twisting of the forging, if done at the correct temperatures, should be no more detrimental than forging the crank in the first place. To support this hypothesis, tests done by a major crankshaft forging company of 200 twisted cranks versus 200 non-twisted cranks indicated that there was no discernible difference in reliability between them.

Pin Configuration

It's easy to make the mistake of thinking that a typical passenger-car V-8 is simply two four-cylinder engines put together at a 90deg. angle. But this proves not to be the case. A V-8 is more like four V-twins bolted back-to-back. If you look along the length of the typical small-block Chevy crank, you will see that all the crankpins are at 90deg. angles to each other. This type of crank is called a two-plane crank. In other words, the big-end journals are situated on two separate planes, and with such a layout the engine can be perfectly balanced.

There is another method of laying out the journals on a V-8—it's called the single-plane crank. With this style of crankshaft the V-8 is indeed two four-cylinder engines, mounted at 90deg. to each other.

Using a single-plane crank appears to have many advantages, at least on paper. First, the crankshaft can be considerably lighter—all other things being equal, it could shed as much as 7lb of weight. Second, the layout of the exhaust system is better.

With a two-plane crankshaft the firing pulses along one bank of cylinders are uneven, but a single-plane crank delivers even firing along each bank. The theory behind the workings of a four-into-one exhaust system is, to a certain extent, predicated on even firing pulses. So a single-plane crank should allow the exhaust system to function better, and therefore lead to the production of more horsepower. A good example of a successful single-planed race engine is the three-liter Cosworth DFV engine which, for over a decade, reigned supreme in Formula One.

On the debit side, a single-plane crank has a balance problem. Anyone who's driven a four-cylinder engine will realize they are not nearly as smooth as a V-8. This is not due solely to the fact that a V-8 has twice as many cylinders to deliver the firing impulses, but that an inline four-cylinder engine cannot be perfectly balanced. Primary out-of-balance forces can be compensated for, but secondary out-of-balance forces cannot. A single-plane crank in a V-8 will cause the engine to shake as if it were two inline four-cylinder engines, so from the point of smooth running, a single-plane crank V-8 is at a disadvantage.

For race applications, however, vibration is not necessarily an overwhelming issue. Over the years, many people have tried to achieve the advantages of a single-plane crank in a small-block Chevy V-8. To date, the effort put into this area of development is only just starting to show a clear-cut supremacy over a two-planed cranked motor. Callies has built several cranks for a well-known engine builder and revision to exhaust systems intake manifold and cams are now starting to show the advantage of a single-plane crank.

Crankshaft Mass and Weight

The sole job of the expanding gases in cylinders is to move mass. The less mass there is to move, the faster a given force developed on the pistons will move it—that's simple physics. Remember that when it comes to high performance, it's acceleration that matters rather than top speed. Thus, taking weight out of the vehicle anywhere helps.

Earlier, we covered how to lighten a stock crank by 5 or 6lb. The question is, was all that effort worth the return? Saving 5lb on a vehicle that weighs 3,000lb may appear somewhat of a wasted effort, but there's more to it than meets the eye.

Most engines operate through a gearbox and a final-drive unit, which means the engine doesn't see the true weight of the car; as a first approximation, it sees the weight of the car divided by the overall gear ratio. By this simple analogy we are saying that if any engine components are lightened, the effect will be proportionate to the reduction in rotating mass compared with what the engine perceives the weight of the car to be, as seen through the gearbox.

For a circle-track car, if the track were relatively short, there could be at least a 10:1 ratio between the engine and wheels. That means the gain is, as a first approximation, that the weight of the car now looks like

The heat-treatment process used on Callies crankshafts is ion nitriding. This type of heat treatment, which takes anything from twenty to forty hours to complete, takes place at relatively low temperatures so that distortion can be avoided. A shot through one of the view ports into ion discharge furnace shows all the parts with an eerie purple glow around them.

Final inspection is done on this Ad Cole automatic checker. This machine keeps a complete record of every dimension on the crankshaft, and measures them to within 0.000004in. Needless to say, the room in which this piece of equipment is housed has to be temperature controlled to close limits.

only 300lb to the engine. Saving 5lb out of the mass starts to look like a more significant amount. Of course as the car's gearing goes up, so the perceived amount of weight saved lessens.

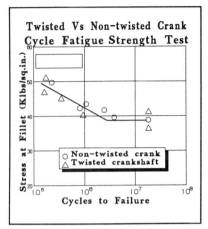

There's been a lot of controversy about twisted and nontwisted cranks. Based on the figures here, there doesn't seem to be a significant difference.

Let's look at this another way. These days, most engines are tested on dynamometers such as a Super-Flow 901, which tests engines under an accelerating mode. Instead of the engine being run at a steady rpm to gather its power, the computer and the various sensors pick up the power as the engine accelerates through its rpm band. As a result, the pressures developed in the cylinder have to expend horsepower accelerating the crank and flywheel of the engine, as well as all its other rotating parts. When the engine reaches a steady speed, power is no longer sapped by the accelerating parts. So under steady speed conditions an engine will show more horsepower than it will when accelerating.

Here's a simple test. We can change the amount of mass the engine has to accelerate simply by changing the flywheel. One test showed that adding 30lb of flywheel weight reduced the engine's output by more than 15hp when it was accelerated at 300rpm per second per second. This acceleration rate is typical of something akin

to about a low third gear, so what we're saying is that it took 15hp to accelerate 30lb of excess flywheel weight into third gear.

The engine's rotating mass isn't solely in the flywheel, a good portion of it is in the crankshaft—although the crankshaft isn't necessarily of the same size diameter as a flywheel; it is significantly heavier on most occasions. Most crankshaft lightening takes place toward the crankshaft's outside radii, which means the effect of lightening is greater than a set of scales reveals.

When we try to accelerate a body rotationally, the mass at the center of rotation hardly moves, so its effect on retarding acceleration is minimal. Mass at the outer edge has to be accelerated rapidly because it has farther to go around the circle, so it

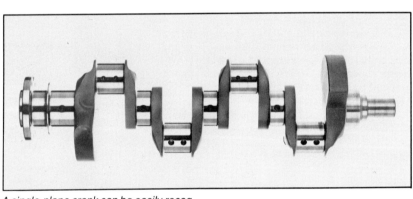

A single-plane crank can be easily recognized by the fact that each big end journal is 180deg. from the adjacent journal.

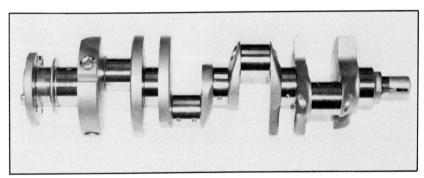

A conventional two-plane crank is used in stock small-block Chevrolets. This has journals at each 90deg. increment around the clock. Other layouts are possible but this is deemed the most convenient by Detroit auto manufacturers.

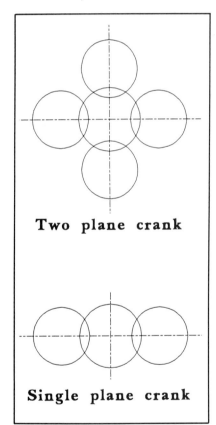

The top sketch is the journal position on a two-plane crank, the common Detroit-style crank. A single-plane crank has the big-end throws situated on one single plane, like a four-cylinder engine. In theory at least, this appears to be better for a race engine because it allows better exhaust tuning, but from the point of view of vibration-free running, it's inferior to the two-plane crank.

has a greater effect on acceleration. Thus mass removed from the outside diameter of a rotating part has far greater effect on acceleration than mass removed in the middle.

So it is with most crankshaft lightening: although the scales reveal a certain poundage reduction, the engine actually sees a larger reduction. But since it is rotating, it is difficult to measure the effect of the reduction.

Rotating Crankshaft Weight

Callies Performance Products has built a device that measures the moment of inertia of a crankshaft. In the chart you can see some figures produced on this device. When you look at these figures there are a few things that you need to take into account. First of all, the amount of the counterbalance weight is directly affected by how much piston and rod weight it has to deal with.

When balancing a small-block Chevy two-plane V-8 crank, you must add a bob weight to the crankshaft throw to simulate the weight of the reciprocating mass, plus twice the rotating mass of one rod assembly. This basically means adding the weight of the piston to the pin end of the rod, and adding double the weight at the big end to this figure. The resulting number is the bob weight required.

From this you can see that weight taken from the big end of the connecting rod has a greater effect on reducing bob weight and thus counterbalance weight than any taken from the piston end. We shouldn't regard this as a means to cut down on our efforts to lighten the piston or the connecting rod at this end, however. Weight taken from the piston means that there's less load in the connecting rod, and therefore the connecting rod will withstand higher rpm.

With a stock crank the typical bob weight, even at its lightest, is around 1,925g and at 52lb a stock cast-iron crankshaft has 0.2918in pounds per second per second of angular inertia. A stock crank modified as shown in the previous chapter, with a 1,552g bob weight, lost weight to the tune of 5lb, a 9 percent reduction; however, its angular moment of inertia dropped to 0.2341in pounds per second per second, a reduction of almost 20 percent.

In other words, though the crank lightened up only 9 percent, the engine actually sees this as a 20 percent weight loss because the mass has been removed mostly at the outer edges of the crank.

When counterbalanced to suit an 1,853lb bob weight, the Callies SuperLite steel crank weighed 52lb and had a moment of inertia of 0.2808in pounds per second per second. As you

may gather from these figures, the bob weight has a considerable effect on the rotating mass of the crank.

Another Callies SuperLite crank, one for a drag-race setup, utilized a bob weight of 1,600g by virtue of a light rod and piston assembly. This crank weighed 49lb overall, but the moment of inertia was down to 0.2357in, so by comparison with a stock crank the engine sees it as some 19 percent lighter to turn. When measured by its weight alone, though, the scales reveal that it's only 6 percent lighter.

To determine how much this weight reduction is worth, we must consider the type of racing. At a superspeedway like Talladega, Alabama, where the car maintains almost its maximum speed throughout an entire lap, the rpm change is minimal so the acceleration rate of the crank is slow: Talladega almost approaches a constant speed. Under these conditions, a lightweight crank is not worth much, to the point of being almost negligible.

On the other hand, on a quarter-mile track, the engine may go through as much as a 4000rpm change in

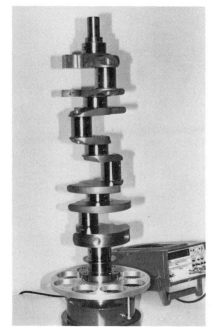

This device at Callies Performance Products is used to measure the moment of inertia a crankshaft has along its center line. Put into simpler words, the moment inertia is the amount of effective weight that the engine sees the crankshaft as having when it has to accelerate it up to rpm.

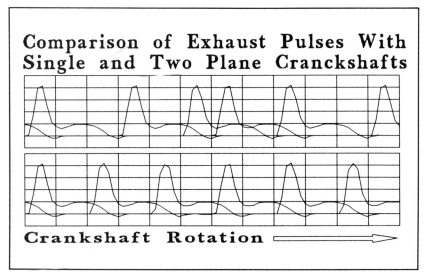

Comparison of Exhaust Pulses With Single and Two Plane Cranckshafts

Crankshaft Rotation ➡

Here are the exhaust pulses from an engine equipped with a single-plane crank (bottom curve) versus one equipped with a two-plane crank (top curve). The single-plane crank will be seen to have even pulses, but the two-plane crank, every two revolutions, has two firing pulses which are only 90deg. apart. This tends to disturb the effectiveness of shock-wave tuning on the exhaust.

engine speed. In this case, the difference can be significant. The engine's acceleration rate is right at, or may even exceed, 1000rpm per second. Under these conditions, lightening the crank by the amount we're dealing with here could free-up probably as much as an additional 10hp to drive the car along, rather than to accelerate the rotating mass in the engine and drivetrain.

As far as drag racing goes, a different situation exists. Often a drag racer will utilize the weight of a flywheel to launch a car, especially if the car is heavy and low gearing cannot be put into it. The success of this technique relies on storing energy into the flywheel and then putting it to use to launch the car.

As any drag racer is aware, the most important part of any drag race is the start—not the finish! The faster the car can build velocity initially, the more effect it has on reducing ETs. Under these conditions a lightweight crank is superfluous because any weight taken out of the crank may be compensated for by the use of a heavier flywheel. However, if the car can be geared correctly so it is launched on its *gearing* rather than on stored energy in the flywheel, then the situation is different.

Most drag-race cars have low overall gearing (numerically high) in first gear. This means that the engine sees the weight of the car significantly less than it really is—it may see as little as one-twentieth of the car's weight. This means whatever weight has been taken out of the crankshaft-flywheel assembly now represents a large proportion of what the engine sees as mass to accelerate.

Indeed, lightening the crankshaft could make the engine think that the car weight has been reduced by as much as 100-150lb in first gear. Of course as the driver shifts up through the gearbox, the apparent weight reduction, as seen by the engine, becomes less. How much this is worth on the drag strip is difficult to say because so many variables affect it. If we took a typical 10sec. car with optimal gearing, the computer program indicates that a lightweight crank could be worth as much as 0.06sec. ET reduction on a car that originally ran 10sec. flat for the quarter.

Crankshaft Inertia Comparison

Crank Type	Bob Weight (g)	Weight (lb)	Angular Inertia (in/lb/sec²)
Stock Cast	1,925	52	0.2918
Lightened Stock	1,552	47	0.2341
Callies Light	1,853	52	0.2808
Callies Super Light	1,600	49	0.2357

Measuring crankshaft inertia takes more than just a set of scales. The scales inevitably deliver the wrong answer—it's not the weight of the crank we need to know, it is the moment of inertia, and as you can see from these figures, a direct relationship with crank weight does not follow.

Lightweight connecting rods, piston assemblies and hollow big-end journals can pay massive dividends in terms of counterweight reduction. Just check out the difference in the counterweight size on these cranks shown here. The one on the left has the stock factory cam-turned counterweight, whereas the one on the right has concentric counterweights. Oil scrapers are more effective on concentric counterweights.

Windage

A significant power loss in the bottom end of the engine is caused by the drag of the crankshaft having to turn through the mass of oil suspended in the air or gathered up by the crank itself. It's easy to assume that because of the high rpm involved, oil will be centrifuged away from the crankshaft onto the walls of the block into the pan.

Actually, this proves not to be the case; the crankshaft is a giant eggbeater, and tends to gather the oil up so that as much as a quart of oil can be going around with the crankshaft. The accepted way to combat this is to add scrappers to the oil pan that knives the oil off the crank and the working parts, cutting viscous drag. Here, several points are worth mentioning.

A cam-turned counterweight cannot have the oil knifed off it as effectively as a circular counterweight (OD concentric with crank centerline). So cutting down bob weights and drilling big-end journals—both of which allow smaller counterweights to be used—can be beneficial since small, concentric counterweights are less prone to windage losses than large, cam-turned counterweights.

We also have to consider the aerodynamics of the rotating parts in the crankcase. At 8500rpm the peripheral speed of the crank is about 90mph. Imagine dragging something through the viscous medium of oil and air at 90mph, and you can see that it's going to absorb a lot more power to drive it if it's of a blunt shape than if it's streamlined.

There's a limit to any streamlining you can do to the bottom end of a small-block Chevy. It's not possible to streamline the rods without going into a major redesign exercise, but to a large extent it is possible to streamline the crank. Knife-edging the leading edge of the crank, at least in theory, should allow the crankshaft to cut through the oil easier and free-up horsepower.

Gary Thompson of ABS has done some tests in conjunction with Chrysler on its small-block motor to determine how much difference leading-edge knife-edging made. Tests were done with three cranks: the first with conventional flat leading edges; the second knife-edged toward the main bearings; and the third knife-edged toward the big-end bearings. Tests indicated that at about 7500rpm, both knife-edged cranks were worth about 4hp. The knife-edged crank that drove the oil toward the cheeks of the main-bearing cap, though it didn't make any more horsepower than the others, aerated the oil less. This test indicates that if you are going to knife-edge the crank, there is an advantage in knife-edging it toward the mains. The actual form of the knife-edging is fairly flexible.

The key to making a knife-edged crank is more a question of taking off the sharp edges. Subsequent tests have indicated that a full radius on the leading edge of the crank achieves the same effect. Indications are that the trailing edge of the crank, however, should be left square so as to shed the oil easier.

This brings us once more to the subject of oil-shedding coatings on the crank and associated parts, a subject we dealt with earlier and worth considering to an even greater extent here. We are now talking about engines that are turning substantial rpm. Pro-Stock engines will be running to 10,000rpm, and shedding the oil effectively at this kind of rpm should result in a measurable increase in horsepower. Consequently, those oil-shedding coatings done by companies such as Swain Industries become even more meaningful.

Oiling System

The stock crank oiling system fares well, evidenced by the fact that stock cranks can go considerable miles without suffering significant wear. But it's not necessarily the best system available.

Before we go into any variations on the stock oiling system, let's consider how the big ends and main bearings are fed by the oiling system. To begin with, the oil pump is only a means of delivering oil to the bearings. It does not provide the pressure to support the oil film between the bearings.

Consider how high the loads are in the main bearings. Some of the main bearings are realizing as much as a 13,000lb load. In plain view there's only about 2½-3sq-in of bearing area to support such a load. Even if the oil is fed in at 100psi, it isn't going to supply enough pressure to support more than about 300lb. To support real-world main bearings and big-end load needs, we need oil pressure as high as 5,000psi to maintain an oil film between the two.

So how does the oiling system work? Basically, it's simple. The oil builds a hydrodynamic pressure as it is dragged between the journal and the bearing. The localized pressure reaches thousands of pounds per square inch and ultimately equals, and so balances out, the pressures exerted by the loads on the piston.

With a better idea of how the bearing works, we can address the subject of where the best point is to feed the oil into the bearing. Should it be fed in at the point of peak loading, or would another position be better?

If an attempt is made to pressurize the bearing with 70psi at the point of peak loading, bearing pressure on the oil hole could possibly limit oil flow out of the hole. On the other hand, if the oil is fed in well away from the point of peak loading, then the oil can feed into whatever gap is there and a dynamic pressure can be built up around the bearing. Putting the oil at an angle of at least 80deg. or more away from the point of peak loading provides the best oil film thickness at the point of peak loading.

An aftermarket crank oiling system such as the Callies system, one of the more highly developed systems on the market, puts oil into the big-end bearing in two places—80deg. and 90deg. from the point of peak gas pressure, and 90-100deg. from the point of peak inertia loading. Not only does this allow the crank to have

Knife-edging the crank in this manner looks like it's worth a little extra in the way of top-end horsepower.

maximum oil film strength thickness throughout the complete two-revolution cycle, but it also gives the biggest scope for putting a large hole in the big-end journal to lighten the crank. If you look at the main bearing, the same kind of technique is applied there—oil is put into the bearing well away from the point of maximum load.

Cross-Drilling

Cross-drilling is a familiar term to many enthusiast engine builders and racers. By definition the term is self-explanatory, but what does cross-drilling achieve?

We know that the big-end bearing sees not only the oil pressure delivered by the pump, but also pressure developed by the centrifugal force due to the rotation of the crankshaft. On a typical 350, the oil pressure at 7500rpm, just prior to the exit point of a big-end bearing, can be an additional 200psi over pump pressure. At high rpm this can actually draw the oil away from the main bearings. In turn, this leads to a starvation of oil at the main bearings.

The remedy is either to increase the oil pump pressure to compensate, or to cut down the flow of oil to the big end at high rpm. An obvious way to reduce high rpm and oil flow is to put restrictor plugs into the ends of the big-end holes and meter the oil, but this then means that at low rpm the big end can be short of oil. With pump pressure down, and the big end restricted, the flow of oil to the big end may simply not be enough.

The solution here is to deliver the oil to the big-end bearing at a smaller radius, as this cuts the amount of centrifugal force seen by the oil. Cross-drilling the crank at the big end puts the oil into the bearing at a smaller radius, and therefore reduces the amount of oil drawn from the mains at high rpm without inhibiting oil flow at low rpm.

Some people also prefer a cross-drilled main bearing on a Chevy crank. The justification for this is simple enough. Oil for the big ends is picked up mostly when the main-bearing drilling is passing over the grooved upper half of the bearing shell. As the crank rotates, the oil hole will pass over the ungrooved lower half. When this happens the amount of oil fed to the big ends becomes restricted.

By cross-drilling the main bearing, the big ends are assured an almost constant supply of high-pressure oil because at least an oiling hole will be passing over the oil-fed groove in the bearing at any one time.

In the past, there has been a tendency for engine builders to groove either the crank or the bottom bearing so as to provide a full time supply of oil to the big ends. This should be considered a definite no-no. Grooving the crank weakens it and a groove in either bearing element reduces the load carrying capability far more than the percent reduction in bearing area would suggest. If more oil is needed at the mains, then drilling the block oil feed oversize and increasing the size of the groove behind the cam bearings can be done.

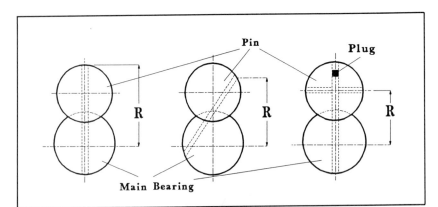

To say the least, the Callies oiling system looks complicated compared with the stock system, however, it will all make sense if you put together what is shown here with what is said in the main text. The A arrows indicate the maximum main bearing load area. These areas develop as a result of maximum gas pressures occurring at points indicated by arrow B and maximum inertia loads indicated by arrows C.

Here are the various styles of main bearing-to-big end drillings showing how they can cut down the radius at which the oil is emitted. The simplest way to drill a crankshaft is as per the left-hand drawing: notice the R is the largest value. A Chevrolet crank is typically drilled as in the center drawing: notice that R has been reduced quite substantially, while still only necessitating a single oil hole. To really chop the radius to a minimum, though, normally requires cross drilling, as seen here on the right-hand drawing.

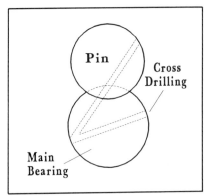

The purpose of cross drilling the main bearing is so that the oil hole feeding the big-end journal can be supplied with oil on a more or less continuous basis. Remember, only the top half of the main bearing shell is grooved.

Crankshaft Preparation

11

Although a forged crank can still be considered the racer's number one choice, cast cranks are suffering from an age-old reputation they hardly deserve. The truth of the matter is, a cast crank can function in performance applications better than most people, even those in the industry, give them credit for.

If you spent a few days with a Magnaflux machine checking crankshafts for cracks you would probably be surprised at what you would find. Such a test would reveal that cracked forged cranks outnumbered cracked cast cranks by approximately 30 to 1. This doesn't mean cast is better than forged. Because of a forged crank's better ductility it can live with a *minor* surface imperfection for a longer period of time before breaking. On the other hand, if a cast crank should develop a crack, it will break almost immediately.

In practice, a cast crank tolerates less fatigue at higher loading levels than a forged steel crank, but tolerates more at lower levels which is why it's more than acceptable for street use. But for short periods a cast crank can cope with the kinds of loads and rpm expected of a high-output race engine.

Rather than overall load it is fatigue that will sooner or later kill a crank, and for a cast crank near its limit, it is usually sooner than later. However, cast cranks can be considerably improved upon, which is good because most crankshafts used in small-block Chevys are cast. If you are prepared to put the work into a cast crank, you can achieve a substantial improvement in its life, power and rpm capabilities.

If you have a forged factory crank, don't feel left out. These same techniques are equally as beneficial but because there is more excess metal, cast cranks have the greatest potential for improvement. If you are going to prep a crank as described here, then the work needs to be done before the final finishing of the journals.

Preparation

First select your crankshaft, preferably a low-mileage crank. Don't use one in which a rod or main bearing has failed and has burned the bearing, since heat cracks can sometimes go deeper than expected.

Once you have a likely candidate, go to your local machine shop and have it Magnafluxed to be sure it is crack-free.

At this point some decisions must be made. The most obvious is, do you want to go all-out on crank preparation, or will a mild prep suit your needs? A mild prep will involve removing any obvious lumps and bumps from the crank and, if you check it out, there are many on a typical cast crank. A mild prep, at best, will only remove about a pound of material, but cleaning up the surface will help fatigue strength and windage.

Drilling Big Ends

If you intend to max out on crank preparation, figure on somewhere in the range of twenty to twenty-five

Here you can see the position of the oil hole in relation to the rest of the crank. The idea is to drill the journal and miss the oil hole.

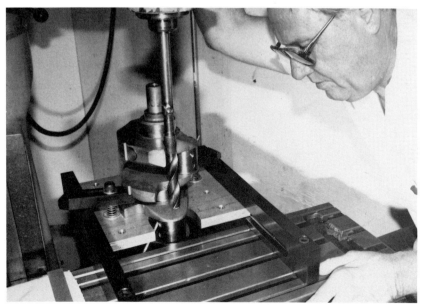

Gary Thompson of ABS is seen here using one of his company's drilling fixtures to drill the big end. The primary purpose of this fixture is for drilling counterbalance weights to install Mallory metal.

This close-up of a big end being drilled shows the position of the drill in relation to the big end. This position is not super critical for a stock casting or forging. All that's needed is to miss the oil hole yet keep the drill sufficiently far away from the surface of the journal as it passes through the big end—0.200in is desirable but 0.150in has, in the past, proven OK.

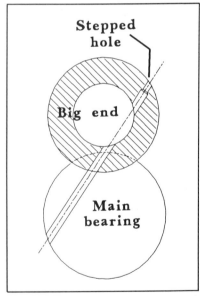

If the big end is to be drilled symmetrically, it will pass through the oil hole. To feed the big end with oil means the use of a tube pressed into the oil hole to get the oil to the big end. The hole needs to be stepped, as seen here, so that the tube cannot come out.

hours' grinding first with a body grinder, then with a die grinder of the normal type used to prep cylinder heads. In addition, the machine shop must drill the big ends. This may sound like a drastic move but correctly done, boring a hole through the big ends actually strengthens the crankshaft, as most of the loads are taken in the top 3/16in of the crank journal.

By drilling the big ends, the opportunity now exists to remove a like amount of metal from the counterbalance weights. This means the counterbalances can either be reduced in diameter, or they can be knife-edged to cut through the oil easier and thus reduce windage losses or a combination of both.

Holes up to about 7/8in can be successfully drilled in the big end of a crank. The easiest procedure is to slightly offset the hole from the center of the big end so it misses the oil drilling. The big end should be drilled to leave at least a 3/16in remaining wall. If you want to maximize the effect the drilling has on the fatigue life of the crank you should drill through the center of the big end, and drill and sleeve the oil hole as necessary from the main-journal end. The redrilling procedure should leave a stepped hole.

Crankshaft Lightening

Drilling four big-end journals with a 7/8in drill removes close to 1 1/2lb of metal from the crank but remember, whatever is removed from the big ends also must be removed from the counterweights. This means the total mass of metal removed from the crank amounts to close to 3lb. You may well ask if the lightening of these parts will help the power output of the vehicle. On a steady-state power test the answer will be no, but the lighter the internal engine parts are, the less power they absorb during acceleration. Consequently, there is more horsepower available at the flywheel to drive the car.

Lightening a crank in this form helps the performance of an optimally geared circle-track, street or drag-race car, all of which are more concerned with acceleration than top speed. If you are building an engine for Bonneville, then lightening the crank will not generate any increase in performance.

If you are building a 400ci motor or an engine with a 400 cast crank in

it—à la 383—then lightening the big ends in this fashion can go a long way toward eliminating any heavy metal from the counterweights to internally balance the engine. If you recall, 400 cranks cannot contain enough of a counterweight to balance a stock-weight rod and piston, so additional weights are added to the flywheel or flex plate and to the damper. This technique gets the engine balanced, but we should ask ourselves whether it's best to balance a crankshaft out on the ends or internally.

As usual, there is no clear-cut answer—although for different situations, different compromises exist. For instance, if you're building a restricted motor that will run at relatively low rpm, say below 6500, then external counterweighting would not necessarily be a disadvantage. In fact, it could be an enhancement because external balancing cuts internal counterbalance size, thus opening up the possibility of reduced windage

Sometimes to clear unavoidable obstructions, it's necessary to approach the journal being drilled with the bit at a slight angle, as is being done here on this 400 crank. Getting access to drill the journal is most prevalent on the intermediate big ends.

losses. It is easier for a crankshaft to turn balance weights in air alone than in a mixture of air and oil.

As rpm increases, however, the addition of counterbalances at each end of the crank will begin to cause an increasingly higher degree of bending approximately at the center main bearing of the crank, thus adding loads to bearings two and four. Any additional loads are resisted by all the other bearings to a degree, but nonetheless, overall bearing friction will increase. With a conventionally balanced crank, bearings two and four are already experiencing the highest loading—the ends of the crank are trying to bend inward. If it's counterbalanced internally, to a large extent this does not happen.

Removing the inner counterbalance weights completely from numbers two and three rod journals is sometimes done to reduce the maximum amount of weight. With these removed, the crank can still be balanced as these two weights oppose each other, though removing them introduces a "couple."

The easiest way to describe a couple is to imagine yourself riding a bicycle. As you push the pedals around, you find that the forward motion of one pedal and the backward motion of the other tend to induce a side-to-side motion as well as the circular path they are meant to move in. The same thing happens to a small-block Chevy V-8 crank unless counterweights are used on the crank to offset this couple. Check a stock crank; to offset the end-to-end couple it has the largest counterweights on the ends of the crankshaft and at opposite sides. The idea is that these weights develop an opposing couple to the one produced by the reciprocating and rotating mass of pistons and rods. As we move toward the center of the crank, the amount of counterbalance weight gets less until the center pair of throws have no counterbalancing.

Crankshafts are produced this way as original equipment for some vehicles, though not Chevrolet. For instance, the 4.9 liter turbo Pontiac engine had no center counterbalance weights on the crank. But this should not be used to argue for such a modification for a Chevy crank; the Pontiac component is not particularly reliable. Nonetheless, many drag racers have successfully removed the center pair of counterweights and made the crank live long enough to get the job done. It is not a crank-lightening technique I would generally employ for engines that must perform for any length of time, however, such as street or circle-track motors.

As for actually removing these counterweights, they can be torched off. This is OK so long as the cut isn't too near any part that mustn't be heat treated. If an oxyacetylene torch isn't

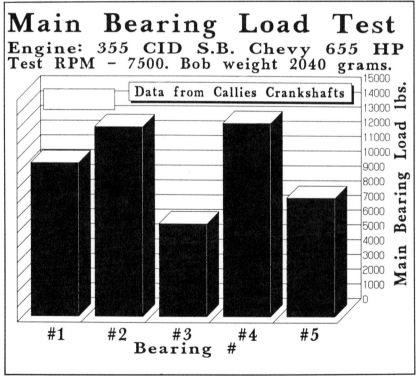

These Callies main bearing load tests in a typical 355ci Winston Cup engine show that the numbers 2 and 4 main bearings can be the most highly loaded. One of the principle problems with strokes of 3.50in or more is that it becomes difficult to put enough weight into the two innermost counterweights. Even on the stock cranks, weight needs to come out of either end counterweights, and put into the number 2 counterweight from each end. This would relieve some of the loads on numbers 2 and 4. The high-speed balancing, as performed at Callies, which spins the crank up to 7000-8000rpm, would tend to indicate a need for extra balance weights at the appropriate position on the crank to counteract these heavier main bearing loads. Basically the direction the industry is taking is to try and cram more weight into those inner counterbalance weights.

The end journals can usually be accessed without any difficulty. In the case of the rear big-end bearing, some metal has to be removed from between the two bolt holes and the flange.

handy, then some industrious effort with a hacksaw or a body grinder will get the job done, though access to a milling machine makes life easier. For the most part, removing the inner counterbalance weights will be a workable alternative only to those prepared to accept increased main-bearing wear. But the modifications about to be described apply to virtually any engine that is required to use a stock forged or cast crank.

Let's assume the first intended modification is to drill the big ends. Up to $^{7}/_{8}$in is acceptable on a big-journal crank. On a forged small-journal crank a $^{5}/_{8}$ to $^{3}/_{4}$in hole should be regarded as the limit. Drilling a 400 crank basically presents little problem because a $^{7}/_{8}$in drill can reach the desired locations.

The second journal from the flange is probably the most difficult to reach. To gain access to the big end, make the area between two of the bolt holes in the flywheel flange concave. The flange modification should be done regardless because it removes metal that is redundant and allows for additional metal to be removed from the counterweight side opposite.

When drilling a 350 crank or any crank with a 3.50in stroke, access to the big-end journals is a little more difficult because other parts of the crank which cannot be removed get in the way. This problem can be eliminated by setting the head of the drill press, or preferably the milling machine, at a slight angle so an extended drill clears any obstructions and still puts the required hole through the journal.

When I prep cranks, I tip the head of the mill 2deg. on each axis, and rotate the ram slide 45deg. from the bed. This allows the drill to come in at a minimum, though sufficient, angle, missing crank projections occurring at some point farther up the drill.

Once the holes have been drilled in the crank the next step is to work with a body grinder, blending the crank and removing excess weight. Rather than try to describe the process at length, refer to the before and after photos to see where metal must be removed and the form it should take. A few noteworthy points: You should strive for rounded surfaces wherever possible. Sharp edges, even external, can be a source of a crack. Also, be sure to polish the internal surfaces of the newly drilled big-end holes, and radius these holes off into the cheeks of the crank. Failure to do this may result in a crank-destroying stress crack.

Crankshaft Counterbalancing

Keep one aspect in mind when lightening your crankshaft: When metal is taken off one side of the crank, an equal amount will need to be taken off the other. It is extremely difficult to estimate accurately, but even a bad guess is better than no guess at all. To get the crank back into balance you will need to work with your friendly crank balancer. Most crankshaft balancing shops will remove metal to balance the crank simply by drilling metal from the counterweight. This gets the crank balanced, but it's by far the least effective way of getting the job done if the intention is to cut windage.

The best way to do this is to use a lathe, taking metal from the OD of the counterbalance weights to decrease the radius and consequently, windage losses. Turning the crank in a lathe as described, however, makes it possible (though unlikely) to be overzealous and actually lighten the counterweights too much. So long as it's not a ridiculous amount, it can be compensated for by drilling into the sides of the counterbalance weights, inserting pieces of tungsten, which is heavier than steel, and welding the hole shut.

This technique is only practical if a relatively minimal amount of extra weight must be added. If the amount is substantial, then it will be necessary to add Mallory metal. This same technique can be used to internally balance a 400 crank.

At this point, to balance the crank you'll need to know the finished weight of a connecting rod and piston because the correct value of bob weight must be attached to each throw to simulate the reciprocating weight of one piston and half the rotating weight of one rod. Probably the biggest amount of material to come off the counterweight of a cast crank will be due to the hole drilled in the big end and will (if you followed the suggested modifications) come off the OD of the counterweight. Even though the crank has been knife-edged, relatively little material will need to come off the trailing edge of the counterweight, since metal removed from other areas has largely compensated.

On a forged crank things are a little different; there is not so much apparent superfluous material toward the

Here is the relative position of the drilled big-end hole in relation to the oil hole.

Once the big end has been drilled, there is so much metal to come off the counter-balance weights that it pays to do the job on a lathe.

center of the crank. When balancing a forged crank, to a greater degree material removal will follow the expected pattern.

On a 400 crank, drilling the big ends might look like a way to achieve internal balance, but this proves not to be the case. Depending on the weight of the rod and piston assembly, you can expect the crank to need one or two sticks of Mallory metal in the counterweights to balance the whole unit internally. Balancing has the intended advantage of reducing main-bearing loads.

Although crank balancing has been discussed here, at this stage it will only be a rough balancing job as the journals have not yet been prepped. The reason all the lightening work should be done on the crank now is that the inevitable slip with the body grinder onto a journal is fixable by the journal regrinding operation. At this stage it's necessary to achieve a reasonable degree of balance to avoid any nasty surprises in the way of large, unexpected amounts of metal removal before final balance.

Journal Preparation

Once the crank is lightened and semi-balanced it is time to think about what needs to be done to the

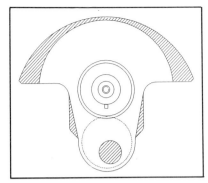

On a 350 crank, I find that going with a lightweight piston and rod assembly and drilling the big ends tends to require the counterweights to be turned about as shown here.

Crankshaft cast iron is really tough compared with that of cylinder heads and blocks. Big cuts just can't be taken. Note the balancing hole welded shut in this crank. If only a small amount of metal is to be removed, the step just above the welded-up hole could also be filled with weld to produce a clean counterbalance weight. In this particular instance, enough had to come off the crank to virtually eliminate the step.

Here's a typical flange on a 350 crank.

After reworking, it loses most of its counterbalance weight.

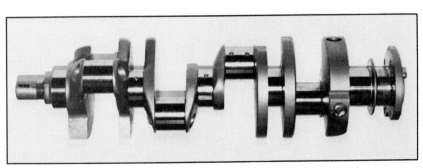

Here's what a crank looks like after all the hand grinding has been done on it. Following the sequences shown in the nearby photos shows you how to get to this point.

journals. The first move is to get the crank ground by a machine shop that has a crank grinder capable of indexing and stroke correcting the big-end journals. Not all crank grinders can do this. Many simply locate and follow off the old pin, and the best you can hope for is that they will average out the errors.

At this point it is best to correct the stroke to a given length; this will save time when it comes to precision

Reworking the flange, it also is undercut as seen here between the flywheel-flex plate bolt holes.

fitting everything together. An accurate stroke length is especially important on a small-block Chevy because the final clearance between piston and cylinder head quench face needs to be narrowly limited if maximum power for any given compression ratio is to be achieved.

The first thing most people worry about when selecting a crank is the size it will clean up at—especially if lots of grinding is required. Such concern stems from the belief that grinding the pins smaller will weaken them.

At first this would seem logical, but cranks do not break through the pins, they break through the webs or through the junction of the web and the pin. The pin itself is one of the stronger parts of the crankshaft; in fact, it's possible to drill a 7/8in hole in it and actually *increase* the crank's fatigue resistance. So, the danger lies not in the grinding itself, but in whether or not a sharp corner exists at the interface between the crank journal diameter and the web.

If the crank must be ground undersize, then one can increase the critical pin-web fillet radius. Making the fillet radius bigger reduces the notch characteristics of the crank. If a fatigue failure is going to occur, then it normally occurs at any change in cross section that looks vaguely like a notch.

Grinding the crankpin undersize and increasing the radius can increase the fatigue resistance of a

crankshaft by a considerable amount —an intelligent guess would be 50-100 percent. Many machine shops grind the cranks with too tight a radius in these corners. The reason is they don't like to radius the corners of their grinding wheels because this means redressing a considerable amount off the wheel for some of the smaller cranks that have a much tighter radius. The grinding wheels are expensive. Having to redress the wheel after grinding a large radius can reduce the wheel as much as grinding five to ten additional crankshafts.

Good machine shops specializing in high-performance crank preparation will keep wheels with large-radius corners especially for such jobs. The bottom line is that if you want the job done properly, expect to pay accordingly. An investment in crank reliability is cash well spent; breaking a crank in the engine will empty your wallet far faster and more effectively.

Journal Sizing

Now for the next hot questions on journals: What size should they be, and how much clearance should the bearings have? As far as the small-block Chevy is concerned, experience indicates that big bearing clearances are not necessarily a good thing. It doesn't make sense to open up the bearing clearances, then have to equip the engine with a high-volume high-pressure pump that saps much more horsepower from the engine than the bigger bearing clearances released. Apart from requiring a bigger pump, big bearing clearances allow a lot of oil to be thrown around in the engine. There is enough power lost through the windage in the oil-laden air in the sump with the stock pump and clearances. Why make it worse?

As if these reasons aren't good enough, the fact is, tighter bearing clearances can support a greater load than loose ones. Unfortunately, tight clearances can generate their own problems. First, the accuracy of the bearing and journal as a pair can become questionable. Although it's entirely practical to generate accurate sizing in the workshop, the considerable stresses endured during operation can cause deformation. Crankshafts bend and connecting-rod big ends can become ovalshaped. The engine's application and the

Here's the stock crank in the region of the rear big-end journal.

clearances selected must reflect these situations.

For a high-performance street motor to be used on a daily basis, the crank should be sized from middle to bottom limit. For a road-race or circle-track motor, up to 0.0005in more can help cut overall internal friction and for a drag-race engine, up to 0.001in extra bearing clearance can be used. The thinking is that for a drag-race engine, oil temperatures don't normally have the time to reach those seen in a road- or circle-track

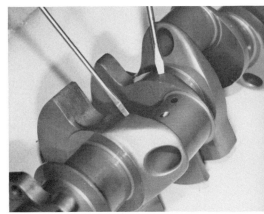

After reworking, they should look like this.

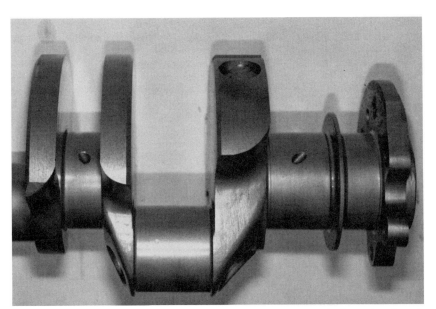

Here's the crank after reworking. Note the knife-edging on the counterweights. This is done in a direction to take the oil away from the rods, pushing it toward the cheek of the main-bearing housings. Many aftermarket cranks are knife-edged so that oil is brought toward the

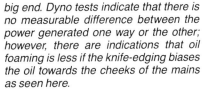

big end. Dyno tests indicate that there is no measurable difference between the power generated one way or the other; however, there are indications that oil foaming is less if the knife-edging biases the oil towards the cheeks of the mains as seen here.

This is the second from rear big-end journal. The screwdrivers indicate the areas from which metal can be removed.

The areas to be reworked are indicated by the screwdrivers.

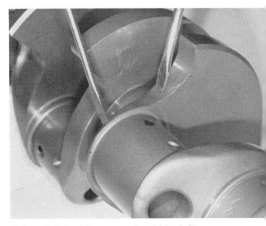

When finished the area should look like this.

109

On the other side of the second from rear big-end journal are some more projections which can be reshaped.

engine, consequently the oil viscosity is a little on the high side and the extra clearance allows for that.

Because crankshaft end float is controlled by thrust faces on the rear main, it's not necessary to let clearances get out of hand. With some crankshafts, especially those that have thrust bearings on the center main, it is often necessary to increase crankshaft end float because crank bending at high rpm can cause clearances to close up. This is not the case with a small-block Chevy.

I recommend retaining standard clearances, and that all clearances be on the tight rather than loose side, as clearances do not significantly change with rpm.

A slightly different situation exists for the side clearance on each pair of rods on the big-end journals. As rpm increases, the crankshaft does bend at this location and the clearances can close.

However, big side clearances in themselves are not recommended. Clearances must be kept to a minimum, just as engine friction must be kept to a minimum at high rpm. Although having an excessively wide clearance isn't going to be the end of the world, it's preferable to keep it down to between 0.010-0.012in. If it's tighter than this, by all means, rub down the sides of the rods to make them fit. If it's looser, then you will

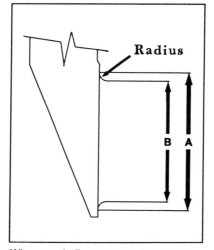

When regrinding a crank, the most important thing is not the journal size you end up with—it really doesn't matter if it's 0.020 or 0.30in under. The most important thing is that the radius in the corner is as big as can be accommodated, because it's at this point that the cranks typically break.

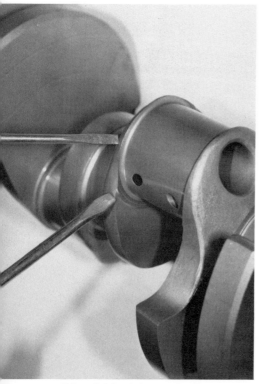

When finished, the area is shaped to the form shown here.

On the front of the crank is quite a bit of scope for metal to be removed, the main mass of it coming from the OD and the leading edge of the front counterweight.

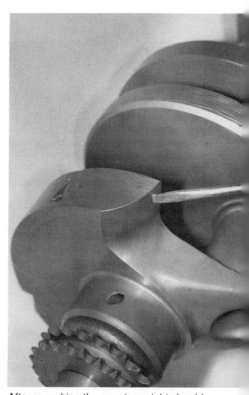

After reworking the counterweight should look like this, but note also, where metal has been removed in the area adjacent to the cheek of the big end. In this shot you can also see how the leading edge of the second counterweight has been reshaped.

either have to live with it or find some means of tightening it. A few thousandths inch can be gained by coating the rods with an oil-shedding finish such as that applied by Swain Industries.

When it comes time to assemble rods to the crank, be sure that an enlarged fillet radius clears the corner of the bearing. Some competition bearings such as those from Sealed Power or Federal Mogul have the corners chamfered to clear larger fillet radii.

Oil Hole Preparation

After the crankpins have been ground, the usual procedure is to prep the oil holes. The technique most often used involves reworking the oil holes with a ball or oval grinding wheel. Possibly the main advantage

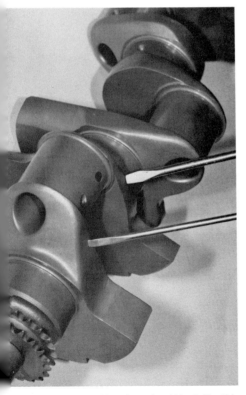

The metal in the areas indicated here are also largely along for the ride.

Material from the points indicated here has to be removed.

These two notches on the web that joins to the center main bearing can be re-formed.

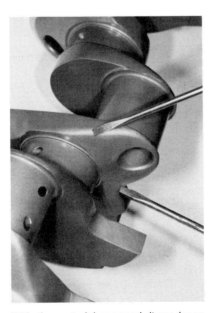

With the material removed, it produces the form shown here.

Here are the re-formed notches.

After reworking they should look like this.

When doing a drastic lightening job as has been described here, it's necessary to make continual reference to the crank's balance so that the final balancing operation doesn't leave the crank in an impossible situation.

of prepping oil holes in this manner is that it is easy to do in a machine shop, and the results look good. Otherwise, the effort seems largely unnecessary, and I'm inclined to put a much smaller chamfer at the exit point of a bearing hole than is common on most crankshafts.

Given the choice, I usually put a small chamfer with a needle file around the bearing hole to make sure the transition between the hole and the journal is smooth, and that there are no metal particles nearby to break away at the sharpest corner of the oil hole-journal intersection. Alternately, the oil holes can be prepped by radiusing the leading edge into the journal diameter. Apart from this, I recommend leaving well enough alone.

Once the oil holes have been prepped, it's time to micro-polish the crank.

Crankshaft Micro-Polishing

This is especially important on a cast crank as it can rectify problems any surface flaws may cause the bearings. If a crank has been ground with a little less than perfect finish, it can be fully corrected by micro-polishing. However, it is important that polishing be in the same direction as the bearing and journal motion. A fine finish is important because a journal with a rough surface can cause heavy bearing wear on initial start-up.

On high-performance engines, the initial start-up and first few seconds of running represent a far higher proportion of the engine's total running time than on a street motor. This and the high rpm involved means demands made on the bearings are greater. Therefore, the bearings should not have any tougher life during the initial start-up than necessary.

Shot Peening

Once the crank journals have been ground, the crank can undergo its final balancing operation. Micro-polishing should be done after balancing so as to remove any marks left by the crank supporting rollers on the balancing machine. After this, the crank should be ready for any surface treatments to enhance fatigue resistance. The simplest choice will, of course, be shot peening which entails masking the crank journals, but leaves all fillet radii exposed. Those corner radii are the most important places for the shot peening to do its job because these are the points subjected to the stresses most likely to break the crank. Do not be tempted to shot peen the crank first and grind the journals afterward to avoid the job of masking. You will defeat the whole purpose of shot peening.

How the crank is reground is important, both in terms of accuracy and reliability. Maximizing fillet radii, even at the expense of crank-pin diameter, reduces the chances of breakage.

Breaks most often occur through the fillet radii of number 4 big-end journals, not through the journal. A big radius at the main bearing's journal is also important. If it takes grinding a crank 0.020-0.030in under to get good corner radii, then do it.

Heat Treating

Though shot peening can be effective, probably a better move is to have the crankshaft heat treated. Cold casing is a relatively common process and is becoming more popular now that the once-popular Tuftriding has fallen foul to environmental protection pressures.

Though I hesitate to say it's the best because of my limited experience with it, ion nitriding looks promising as a means of increasing crankshaft strength. Whereas the Yellow Pages will almost certainly find you a company that can do cold casing, ion nitriding is, at present, a little more uncommon and it will probably be necessary to send away your crankshaft to one of the few companies that specialize in this form of heat treatment. Nitron, in Massachusetts, is one.

Oil-Shedding Coatings

While on the subject of surface treatments, if you really want to put the icing on the cake, as a final operation the crank can be finished with an oil-shedding PTFE-type coating. Such coatings are, in simple terms, a little like those used on cooking utensils. Instead of oil clinging to the surface, it has a tendency to stay in droplet form and as a result is centrifuged off easier. I've never done a back-to-back test of a coated and non-coated crank, but most top Winston Cup Grand National engines are using coated cranks. Though no one I've come across wants to confirm it, rumor has it that these coatings are worth 2-4hp at the top end of the rev range, which for a typical Grand National-style engine is approaching 8000rpm. Bearing in mind that anything related to crankcase windage tends to follow a square law, a coating worth 4hp at 8000rpm quickly drops as the engine's maximum rpm comes

down the scale. For instance, at 6500rpm the gain from an equally effective coating would be about 2hp. Still, every little bit helps and if your budget can stand it I recommend having the crank coated by a company such as Swain Industries or Polydyne Coatings.

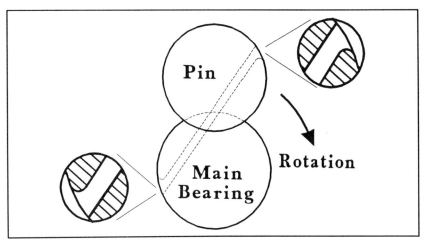

As of 1991, this technique for blending the oil hole into the journal is considered to be the most effective. Tests at Chevrolet have shown a considerable increase in big-end pressure when the oil discharge is modified in this manner.

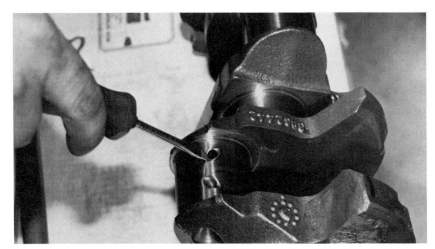

Chamfering the oil holes in the rod journals, as seen here, is a common procedure. In practice there are plenty of indications that this is largely unnecessary, and simply lightly chamfering the hole with a needle file will suffice.

12

Bearings Selection and Preparation

Selecting and installing the bearings for any engine is important, but for a high-output engine the type of bearings used and how they are installed is critical to the engine's life.

Bearing Materials

Gas and inertial forces in a high-output small-block Chevy engine can load big-end bearings to unit pressures in excess of 20,000psi. Up until just before World War II the common bearing material used in most internal-combustion engines was white metal. The technique was to cast this into the rod or the main bearing and then machine it in place. This material is inadequate for a high-output engine, as was anticipated by the aircraft industry just prior to the war.

It became evident that aircraft engines could be the deciding factor in an international conflict, so the development of high-output aircraft engines was spurred on by the need for air superiority. Bearings became a problem early on, and this led to the development of the modern tri-metal bearing. At first, the bearings used in the supercharged aircraft engines employed combinations of silver, aluminum bronze and steel.

The auto industry soon followed suit, but obviously, using rare metals for bearings proved impractical in such a cost-conscious market. Apparently, the first company to develop a successful tri-metal bearing was Clevite. Introduced as the Clevite 77, the bearing set the pattern for virtually all subsequent bearings irrespective of manufacturer.

If you look at the function of a plain metal bearing pressure-fed with oil, you may question why the material of the bearing is important; after all, if the two metals never touch due to the oil film, why would the metal characteristics be important?

There are several reasons why the bearing construction is important, especially at the load levels seen in a modern high-output engine. Not only are the pressures exerted on the bearing high, but they are also cyclic. This means that a material too soft and low in strength can fatigue and crack. On the other hand, if the material is too hard, any foreign particles that find their way between the bearing and journal will cut both to ribbons. Thus the bearing needs to be stiff enough to support the loads, yet have a soft enough surface to allow particle embedment. This is exactly what the modern tri-metal bearing seeks to achieve.

There are many types of metals used in current bearings, and technology is constantly improving. The principal alloys used for tri-metal bearings are steel, for the backing, which is almost universal. For the intermediate level, various bronze and aluminum alloys are common. For the soft, embedable layer—which may only be a thousandth inch thick—a variety of alloys are used: these can contain proportions of aluminum, tin, copper, lead and indium, plus trace elements.

For various applications, manufacturers often produce different grades of bearings. Though by no means universal, a trend is that the higher the rpm and output, the stiffer the bearing has to be to handle the loads. To a degree, this may mean sacrificing some of the embedable layer.

Chevrolet Morraine bearings are available in a number of different grades. Among those types that have seen common use are the Morraine 300, 400, 420 and, a few years back, the 500 grade, which is now discon-

This supercharged 350 Chevy is one of several that have been built and dyno tested in my shop. Horsepower figures are typically around 600, with torque usually being between 525-550lb-ft. This puts a lot of load on the bearings, but when everything's done right it all works.

tinued. My experience with the Morraine bearings has been mostly with the 400 grade, and these I've found to give good results in a wide range of high-performance engines, from street to all-out race.

Another bearing I highly recommend is the bearing originally known as the Clevite 77. This company was taken over by Michigan Bearings in the late 1980s, therefore this grade of bearing is now known as a Michigan 77. Additionally, the TRW and Seal Power bearings can be counted on for good service, as I understand these are basically a Clevite 77 type bearing.

Another brand I've used successfully in high-performance street motors is the Federal Mogul bearings. In the late 1980s or early 1990s, Federal Mogul made some bearing design changes which appear to have made them more able to withstand the rigors of high cylinder pressures.

Another bearing type worth mentioning is produced by Vandervell. This company manufactures a unique lead indium bearing which has proved successful in the high-output Cosworth Formula One V-8s. These V-8s are turning 11,000 plus rpm for long-distance races. The Vandervell bearings have held up under these conditions with virtually no problems. Although I haven't used them on any high-output Chevy engines, I have heard that some of the leading Winston Cup engine builders have used them with great success.

Bearing Crush

The first thing to understand is that retention of a typical rod or main bearing insert is not by means of the bearing tang, but by virtue of bearing crush as the cap is assembled to its mating component. The tangs simply locate the bearings in the correct position for assembly, and it is crush that holds the bearing in place.

If you are building a high-performance Chevy engine it is advisable to set the housing size of both rod and block to the minimum specified to utilize the maximum bearing crush. For the rods, especially stock rods, this is important because even with brief detonation it is possible to distort the big-end housing. This distortion can momentarily overload or remove the oil film so that metal-to-metal contact takes place. The high friction generated can, in turn, cause a spun bearing. A tightly clamped

bearing is less likely to do this. Aftermarket rods are far less critical in this area because of greater support around the big-end bore.

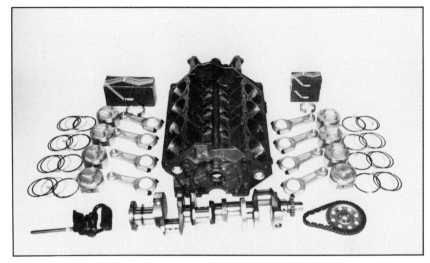

This is how a complete Pro Stock bottom end looks when it goes together as new. Successfully applied assembly techniques, especially in the bearing area, will ensure that it comes apart looking much the same.

For the block, a tight fit between the bearing and housing is mandatory to encourage good heat conductivity. Although heat conductivity is not super-

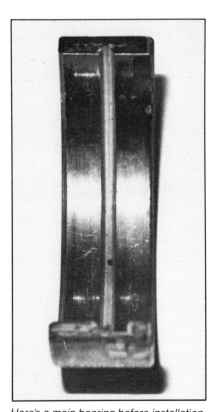

Here's a main bearing before installation. If it looks like it's been polished clean, that's because it has. Thorough cleaning of both the back and front of the bearing is of paramount importance.

Accuracy at the bearing housing is a must. Honing is fine for both rods and main bearing housings.

critical, it may just prove to be the saving grace under less than normal circumstances that may have caused the oil to reach excessive temperatures.

Somewhere between 300-350deg. F., most modern mineral oils begin to break down. The temperatures reached in the bearings are far higher than the bulk temperatures seen in the pan, and under extreme conditions some pretty wild things can happen between the bearing and the journal.

If the bulk oil temperature gets too high, it is possible for the added temperature in the bearing to allow the oil to actually flash to a vapor. If it's going to occur, it usually does so as the piston goes over TDC and the load is taken off the bearing, thus reducing the oil pressure between the top half of the rod and the journal. Allowing the bearing to dissipate as much heat as possible under marginal conditions may just make the difference in the bearing's survival.

Surface Finish and Alignment

For all practical purposes, honing the big ends is the simplest and easiest way to generate the required bearing fit. As far as the main-bearing housings are concerned, two options exist: line honing or line boring. A finely honed finish will produce a surface more able to conduct heat from the bearing; however, a finely bored finish can be acceptable.

Unless an extreme case is considered, more emphasis should be put on the bearing crush than the type of finish. Getting the big-end rod-bearing bore square to the sides of the rod and parallel to the pin bore is *extremely* important. Any error here will accentuate the tendency to knock wrist-pin retaining clips out of the piston pin bores.

As for the block, alignment of the main bearings is not as critical as many race engine builders like to believe. Industry tests done on a low-output four-cylinder three-main-bearing engine—where crank friction quickly shows as a major percentage loss in power—demonstrate that little or no measurable difference in output was seen until the center main was more than 0.003in out of line. Tests on V-8s have shown similar results.

Just how much difference aligning the main bearings makes often seems academic, since the forces of combustion can bend and twist the block more than 0.006in out of alignment from the center main to the block's extremities. If the block can induce this sort of misalignment by virtue of internally generated forces, then one might expect that deliberate counteracting misalignment to reduce high-speed alignment should result in more power.

Attempts to do this with different-thickness bearings do not appear to have shown any measurable power increase. At first this seems to defy logic. However, if we introduce two factors, the picture does clear up, to an extent.

To start with, if the block is bending, then the crank must also be bending, so that any misalignment over the length of one bearing is minimal—certainly far less than the clearance.

The second factor is that a properly developed oil film between the bearing and the journal can sustain loads in excess of 20,000lb on the size of bearing typically seen in a small-block Chevy. In other words, it will take a load of over 20,000lb to break down that oil film, and as long as the film remains, the friction between the shaft and the bearing is low. So, although the crankshaft is bent, the effect of that bend on the friction level in the engine is minimal.

Flow Versus Temperature

Although it may seem straightforward, there is quite a delicate balancing act between the viscosity of the oil, its temperature, and bearing clearances. Thick oil absorbs too much power, especially if bearing clearances are tight; consequently, these clearances have to be increased. More horsepower is lost to windage and pumping the oil around the engine as a result.

Thin or low-viscosity oils are easiest to pump through the engine and they absorb less power, but they also generate a thinner film. Therefore, the accuracy and stiffness of parts plus the quality of the finish have to be much better, and there is less room for error.

If clearances are too tight the oil temperature will climb too high, and the oil film will break down. If clearances are too loose there will be too much oil flying around in the engine, and power will be lost due to viscous drag. Apart from that, big clearances are less able to support the high loads generated in an engine. Tight clearances can support high loads so long as the oil temperature doesn't go out of sight.

When honing the housing be careful so not too much metal is removed. If new main bearing caps are installed, they will need to be line-bored.

With all these variables you can see that there can be a whole range of choices. Fortunately, a great deal of work has been done in this area, so there is no real mystery as to what is required in the way of clearances in the bottom end of a small-block Chevy.

Main Bearing Clearance

A small-block Chevy can be put to many applications—anything from a high-performance, but totally street-able motor, to a Pro-Stock-style race engine. Each may require something a little different. If you're intent on building a refined motor that runs quietly for the street yet has reasonable output, then think in terms of a minimum of 0.002in main-bearing clearance. You can be a little wider than that, but certainly do not go tighter than 0.002in. If it's a regular street motor you are building, where silent running above and beyond anything that a stock motor does is not important, then 0.0025in main-bearing clearance will work just fine.

For a high-output engine, clearance can go up to 0.003in; if it's a high rpm engine, such as found in drag racing, then 0.0035in will be the limit. Except for the lower limit, these bearing clearances are not highly critical; 0.0002 or 0.0003in under to 0.0005in over won't make a significant difference.

As far as end play on the crankshaft is concerned, I find that because the clearance is controlled on the rear bearing there is no need to have excessive end clearances, like those engines that have the end play controlled on the center bearing. When end play is controlled on the rear bearing, the effect of crankshaft flex tightening these clearances is minimal, so there is no need to make an allowance for it.

For a street motor that is required to run as silently as possible, I've used clearances as narrow as 0.005in, but even for high-performance applications, I've found no real need to go wider than 0.007-0.010in.

If a heavy clutch is used, then it's possible that the thrust bearing may not be up to taking the throw-out loads. To combat this, it's practical to pressure feed the back face of the thrust bearing. This is achieved simply by chamfering the edges of the top rear bearing along the split line from the groove to the thrust bearing face. The chamfer need not be large—about 0.030in is usually enough. This bleeds off oil from the groove and feeds it to the back face, keeping a good film of oil between the bearing and the thrust face of the crank.

Rod Bearing Clearance

In general, rod bearings need to be a little tighter than main bearings while still observing about 0.002in minimum clearance. This applies for those applications where minimum crank rumble and noise is sought. For most situations, street or otherwise, 0.0025in will work just fine. For extremely high rpm applications this can be widened to 0.003in.

If the engine is using aluminum rods the clearances required are different. The amount of crush on aluminum rods is higher and this can close up the clearances. As the rod warms and expands some of the crush is relieved, allowing the clearance to increase. Consult the rod manufacturer as to the recommended clearance, but you can normally expect the clearance on an aluminum rod to be about 0.0005in less than would be used on a steel rod.

Another point you should note with aluminum rods is that the lower bearing shell needs to be pinned in place. Again, this is partially due to the fact that some bearing crush is lost when the rod expands. The pin in the cap half of the rod acts as an anchor to hold the bearing in place. Some bearing manufacturers such as Federal Mogul and Sealed Power market bearings that have the requisite hole drilled for the pin location.

The rod side clearance used must reflect the applications you have in mind for the engine. For a low-rpm engine, side clearances are dictated more by machining tolerances than crank flex taking up the clearances and tightening on the rods. As rpm and horsepower go up, however, the need to look at side clearances becomes more critical as crank flex can absorb some of the side clearance.

For a 5500rpm engine a side clearance of 0.010in is all that is necessary with a stock connecting rod and stock-style crankshaft. For a high-performance street-style engine up to a high-rpm Winston Cup-style engine, clearances may range from 0.012in to as much as 0.016in, the small-journal cranks requiring a little more clearance than the inherently stiffer large-journal cranks.

If aluminum rods are installed, the cold side clearance needs to be between 0.018–0.020in. When the en-

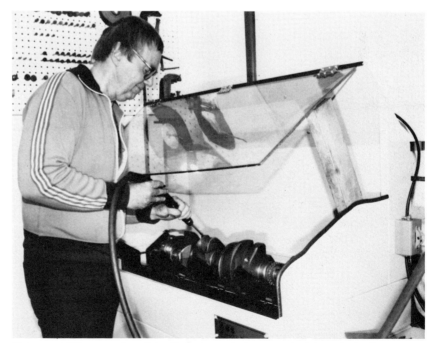

Using an ABS vacuum bench to eliminate the spread of grinding grit around the workshop helps engine reliability. It only takes one particle of grit behind a main bearing to cause a problem. The crank being reworked here is having excess material removed to lighten it. This grit can easily stay in the oil way, so it's important to clean out the oil ways.

After all the main bearings are located in the block and the caps, they need to be lubricated with a suitable engine building mix.

Some oil seals may leak because the crank has worn a groove from previous use. If so, you will be interested in this range of seals from Engine Tech. One is a stock seal for cranks in good condition, the other has a more aggressive preload on the crank to make up for a slightly grooved crank, while the third has the lip moved about 0.050in back so that it bears on an unworn part of the crank to produce an effective seal.

After lubing and installing the crank, each main bearing cap should be torqued down separately, and the crank checked to see if it spins freely.

gine is hot, these clearances end up about the same as for a steel-rodded engine.

Bearing Installation

Installing bearings into a high-performance small-block Chevrolet involves more than just unpacking them from the box and positioning them in place.

First, each bearing needs to be inspected, and since different manufacturers use different packing techniques, the first step is to clean the bearings. Some bearings are packed in grease, others are not. If the bearings are greased, fast-evaporating paint thinner is good for cleaning them.

Check to see that the bearings are the right size for the crank (this is a preliminary check if you have a 10/10 crank). Then make sure the bearings are the corresponding undersize. Next, take a fine needle file or a flat stone and run it over the numbers on the back of each bearing to make sure they are not significantly raised, so as to prevent seating snugly into the housing.

Then, check both big ends and mains on the crankshaft to see that the chamfer clears the radius in the corner of the crank journals. If the crank has been ground undersize by a grinder who's concerned about the crankshaft staying in one piece, the corner radii could be larger than stock. If this is so, the bearings, especially the big-end bearings, may need to have the side chamfer increased to clear this radius, but do not chamfer the bearings any more than necessary to clear the crank radius.

The final check that the chamfer is adequate involves installing the bearings in their housings and ensuring that they do not ride on the radii. This is especially important on the connecting rods.

Prior to installing the bearings in the rods or caps, wipe the housing surface clean with an absorbent paper towel to make sure that it is totally grit-free. If the block or the rod has been previously used, then ensure that the surface is clean down to the metal by rubbing lightly with a 3M Scotch-Brite pad, or similar product, and then wiping clean.

Bearing Preparation

There are several schools of thought as to how the bearing surface should be prepared. The back of the bearing shell should be as clear of oil as possible, so it needs to be thoroughly wiped with a clean shop rag. Depending on the type of bearing, the actual bearing face can be cleaned off with a Scotch-Brite pad or polished with a clean paper shop towel. Which technique you should use often depends on the type of bearing involved.

The Morraine bearings seem to be cleaned best by using a well-worn Scotch-Brite pad. Remember, though, this is not a metal removal exercise so much as a polishing exercise. With Vandervell, Michigan 77, TRW and Seal Power bearings, I prefer to give them a good rubbing over with a clean shop cloth until the bearing surface is shiny. For the latest Federal Mogul bearings, a light treatment with a Scotch-Brite pad seems to be in order.

When the bearing housing and shells have been scrupulously cleaned and the shells installed in the block or rods, some pre-lube must be applied. For this, I recommend Michigan Bearing Guard or a 50:50 mixture of Crane Cam Lube and quality multiweight motor oil. I use this combination because it sticks to the bearings.

If the engine is to be left for a long period of time, I prefer the Michigan Bearing Guard, as it stays on the bearings indefinitely. However, if the engine is to be fired up within a few weeks of it being built, then the Crane Cam Lube and oil mix works just fine.

These recommendations are for the final assembly. If the engine is going together on what is often termed a dry build, then I use a lightweight oil to prevent scuffing during assembly turning torque tests.

Cam Bearings

There are no real concerns with Chevrolet camshaft bearings for most street and drag-race use, so long as the bearings are reasonably in line. To test for this, see that the cam can be installed and rotated freely.

Often, too much emphasis is put on installing the cam without nicking the bearings as the cam is inserted into the block. I've seen cam tests involving as many as a hundred cams going in and out of the block, leave the bearings looking like they've been through World War III! In spite of this, there was no apparent drop in oil pressure, or any other problems.

If you patronize a top-notch machine shop, you'll find that most instances of a stiff crankshaft are caused by dirt under the bearing because you didn't clean well enough. Don't be afraid to use those white shop towels. Grinding grit is immensely more destructive than a few strands of paper particles.

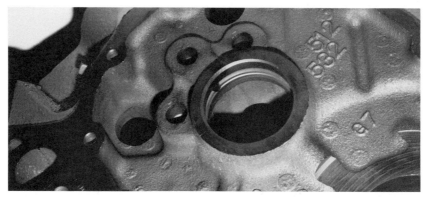

If damage to the cam bearings is to be avoided, they should be installed with a close-fitting bearing driver centralized in the cam bores.

Don't even think about assembling the rod and piston onto the big end unless you're using these bolt sleeves to protect the big-end journal.

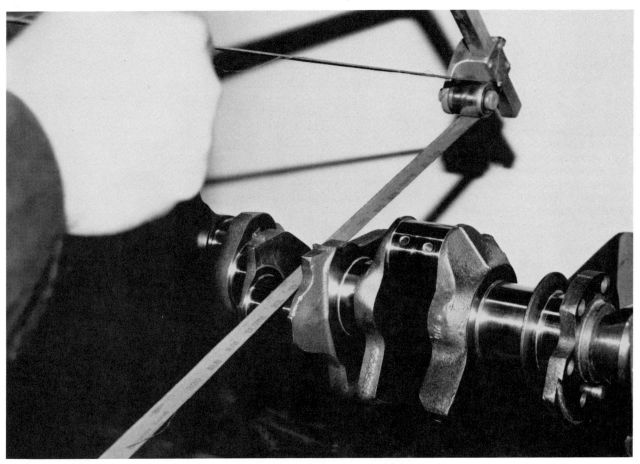

Both the polishing and the method used to achieve it are important for the life of a crankshaft and bearing combination. Not only must the finish be fine, but also the crank must be rotated in the direction it does in the engine, with the polishing agent going the opposite way.

Obviously, nicking cam bearings with cam lobes is not as critical as it is made out to be.

A situation where some cam bearing problems can exist is with endurance-race engines competing in typical 400 mile events. Here, good alignment and high-quality oil is the simplest safeguard against failure. If funds allow, an excellent alternative is to use the Pontiac version of the small-block Chevy, as this is equipped with a set of roller-cam bearings.

Bearing Journals

The objective of all the prep work described so far is to keep the bearing intact to ensure its survival. Since the bearing shell represents only half of the picture, there are some additional factors to consider. No matter how well you prepare the bearings, if the journal running in that bearing is not equally well prepared, the bearing will fail.

Probably the most important aspect of any plain bearing lubricated by a hydrodynamically generated oil film, is that the journal must be round. If it is lobe shaped it will quickly destroy the bearing.

The surface finish must also be as smooth as possible. A finish produced by a grinding wheel, no matter how fine, is not good enough. The journal must be polished after it is ground.

You must also ensure that the correct clearances are adhered to. Measuring the clearance on a bearing involves measuring the vertical clearance, that is, clearance at 90deg. to the split line. The way bearing shells are manufactured, more clearance exists at the split line than at the vertical point. This is done deliberately so that as distortion pulls the sides of the bearing in, the shell doesn't act as a scraper removing oil from the journal. In a professional engine-building shop, special and expensive micrometers will be used to measure both journals and bearing bores.

Building an engine at home, you are unlikely to be able to afford this equipment just for the few engines that will be produced. Probably the best way to determine bearing clearance under these conditions is to use the plastic gauge system available from Sealed Power dealers. The technique involves using a thin, accurately sized strand of pliable plastic. This is inserted between the bearing shell and journal, and the securing bolts are tightened. The width of the now-flattened plastic is checked against a chart which shows width versus clearance. The wider the deformed plastic is, the less clearance there is.

Bottom End Balancing

13

The object of balancing any internal-combustion engine is to use the inertially developed forces in one part of the engine to counteract those in another part. If this is done successfully, the engine can be completely balanced, and it will run smoothly without undue vibration—indeed, except for firing pulses it should be able to run as smoothly as a turbine. However, some piston, crank and cylinder configurations are not physically able to deliver complete balance.

Goal of Balancing

The main aspect to consider with a piston engine is that there are two principal forces that need to be balanced out. These are known as the primary force and the secondary force.

The primary force is easy to understand; it is force generated by the piston moving up and down the bore. For a moment, let's consider half of a V-8. Assume we have a four-cylinder engine, and that the crankshaft has the two center throws at 180deg. to the two end throws. While the center two pistons are moving up toward the top of their stroke, the outer two pistons are coming down. The inertia loads of the outer pistons balance out the inertia loads of the inner two pistons, and this is called primary balance.

Life is not always as simple as we would like it to be, however. Checking the motion of a connecting rod and piston in relation to the crank reveals that the crank turns a greater number of degrees at the bottom of the stroke to push the piston from BDC to halfway up, than it does from the top of the stroke to bring the piston halfway down the stroke. This means that the motion of the pistons traveling toward the bottom of the stroke is not exactly the same as the motion of the pistons at the top. This being the case, we cannot expect piston assemblies moving around the bottom of the stroke to exactly counterbalance those near the top because acceleration forces will be slightly different.

The difference in these acceleration forces is called secondary out-of-balance in a four-cylinder engine.

Counterweighting

Counterweighting can never completely negate the effect of secondary out-of-balance forces. Centrifugally developed force produced by a crankshaft counterweight is constant so can only provide an opposing force to the continually changing reciprocating forces at two points per crank revolution. What is the result? A four-cylinder engine can never be totally balanced.

Primary and secondary forces in a V-8 can be, but it depends on the crank configuration.

This subject was touched upon earlier when I pointed out that a single-plane four-cylinder-style crank can never be fully balanced in a V-8. That is because the V-8 is now simulating two four-cylinder engines 90deg. apart. As a result, it suffers vibration levels almost equal to that of two four-cylinder engines.

On the other hand, a two-plane crank, with each throw at 90deg. to its neighbor, can have primary and secondary balance, but unless suitable counterweights are put on the crankshaft, it will have a couple.

More detailed analysis reveals the possibility of using a completely non-counterbalanced crank, and containing all the counterbalance mass required in the flywheel and damper. This means any couples or forces that are required to be counterbalanced must be transmitted along the crankshaft to be balanced at the ends. In other words, all the counterbalancing forces are developed at the ends of the crank.

The forces that put the crank out of balance in the first place occur, however, at even intervals along the crankshaft, namely, at each big-end journal. If no effort is made to counterbalance the out-of-balance forces near the point they are produced, the main bearing closest to that out-of-balance force has to take excessive loads. For that reason, internally

balanced engines are better than externally balanced engines.

As discussed, often it is necessary to compromise, and if rpm levels are not too excessive, an externally balanced crank can do the job, especially if it allows the crankshaft to be considerably lightened. For a long-distance race engine, though, it is necessary to minimize bearing force loads. So, for an engine that has to sustain high rpm, and may not change in rpm significantly from the slowest turn to the fastest straight, an internally balanced crankshaft is better. Not only should the crank be internally balanced, but it should be balanced along the entire length of the crank.

Counterbalance weights for the center pair of throws are hardly worth the effort of putting them on; counterbalance weights a little farther out from here are justified, however. Modern balancing machines can only pick up out-of-balance forces as they are transmitted to the ends of the crank, so any balancing that is done to the crank, by virtue of the counterweights' spacing, has to be estimated and finally rectified by drilling on the ends of the crank.

Practicalities of Balancing

When you take your small-block Chevy crank to be balanced, the machine shop doing the balancing will need the following items: a piston, a set of rings, a connecting rod, a wrist pin and a set of bearings. They will also need the crankshaft damper and the flex plate or flywheel. The crankshaft dowel pin should already be installed in the crank.

The technique used to balance the crank will entail weighing the piston assembly along with the rod small end and big end. To balance the crank, bob weights must be added to the big-end journals. The amount of bob weight required will equal the weight of one piston, plus the reciprocating weight of the connecting rod (the small end), plus twice the rotating

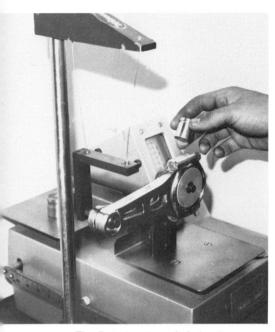

The first step towards balancing an engine involves measuring the connecting rods, big-end and little-end weight. To the little-end weight, the weight of the piston must be added.

To balance a two-plane crank, it's necessary to add to the big-end journal the reciprocating weight of one connecting rod and piston assembly, plus the rotating weight of two.

weight of the rod (the big end). An allowance of about 2g is also made for oil. With the requisite weight made up in the form of a bob weight, they are installed onto the crank throws, and the crank is then rotated. Holes are drilled in the counterweights until the crank spins with no out-of-balance reading.

What has just been described represents a conventional balancing job. If you have lightened the rods and pistons considerably, but have done little work to the crank, you will need to do some serious drilling in the crank to get the counterbalance weights down.

Drilling, however, is the *worst* way to lighten the counterweights when a lot of material has to be removed. What this does is create a lightweight counterweight that has all of the internal drag of a large counterweight.

As mentioned earlier, any excess material on the counterweights should be turned off in a lathe. In many cases the amount of counterbalance weight on the crank is inadequate for the weight of piston and rod assembly being used. This often occurs when a heavy-duty piston and a steel endurance-type rod such as a Carrillo is used. In that case, it's necessary to add Mallory metal to a stock crankshaft to achieve balance. Mallory metal is virtually as heavy as lead. It's inserted into the crank to

increase the effective counterbalance weight.

If the crank has to be balanced this way, it's expensive. One of the advantages of buying an aftermarket crank and rods is that the crank manufacturer will make the crank counterbalance weights an appropriate size to achieve balance without necessarily adding Mallory metal.

If you patronize a good machine shop, they'll do a fairly accurate job of balancing your crank and assembly. But it's worth asking just how important it is to get the balance spot on, considering that when the engine is balanced, the parts are rotating in air and they are all clean, but in use they get doused and pick up quite a bit of oil. Carbon builds up on the top of the pistons, and though the carbon may be light, the amount of oil that can splash on the crank and onto the rods and undersides of the pistons can vary greatly. As I've said before, at high rpm the oil can become entrained in the crankshaft. It is most unlikely that it gets entrained in any form that is completely balanced, so that can throw off the balance of the crank.

There are other factors to consider. For instance, since the atmosphere in the crankcase consists of a mixture of oil and air, we can say that the average density of this mixture in the case is substantially higher than that of air. As the piston accelerates up the bore it is likely to accelerate away from droplets of oil, and any that are on the piston are likely to be shed. That means a piston traveling *up* the bore tends to get lighter. On the other hand, piston traveling *down* the bore tends to get heavier because it collects oil both out of the atmosphere and off the walls of the cylinders.

Now we have a situation whereby pistons going up the bore are lighter than those coming down—an artificial imbalance has been created. How much of an imbalance would be almost impossible to tell, and certainly, it's unlikely to be consistent from one engine revolution to another. So the question remains, just how accurate should the balance be for smooth running of the engine? Do the bob weights need to be totally accurate, or should we compensate for excess oil on the moving parts.

Overbalancing

Some engine builders specify that the cranks need to be overbalanced. To accomplish this, the crankshaft

The Mallory metal slugs are installed like this. One way or another it's quite expensive.

A less-expensive technique that's especially useful for 383s and 400s is to add some external counterbalancing in the form of this ABS counterweight plate that fits between the crank and flywheel. This plate allows non-external balance components to be used.

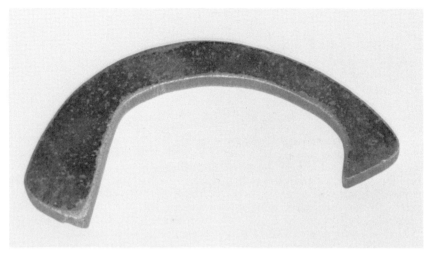

Here's another product from ABS, a weld-on cheek counterweight for 400 cranks. So long as the rod and piston assembly is not really overweight, these cheek counterweights will avoid the necessity of Mallory metal.

bob weight is increased by the odd 1 or 2 percent over the normally calculated weight, which means that the crank counterweights are slightly heavier than those on a neutrally balanced crankshaft.

According to some industry experts, a slightly overbalanced crank runs smoother in the kind of rpm range expected of an all-out race engine, that is, from 5000 to maybe 8500rpm. On the other hand, the neutrally balanced crankshaft is better for applications that are going to run from 1000-5500rpm, maybe 6000 max. I have not been able to prove this one way or the other. As of 1991 no instrumented test engine has been run to determine if overbalancing actually cuts high-speed vibrations.

There is another aspect to overbalancing pointed out by Grumpy Jenkins that does appear to be useful, however—especially if you are building a number of engines and need to build up a stock of instantly interchangeable parts. Piston and rod assemblies rarely weigh exactly the same. If you are going to mix different weights of pistons and rods so that the engine isn't exactly balanced, then it is better to have it overbalanced than underbalanced.

In other words, if you know what is the lightest and heaviest piston and rod assembly likely to be used, and balance for the heaviest one, within reason, it will work for the lightest one as well. We assume, of course, that we are not talking about much more than 20-30g difference between the lightest and heaviest rod and piston assembly. By keeping the weights of the reciprocating parts within reasonable bounds this does mean that cranks, rods and pistons can be made interchangeable between several engines.

Vibration Dampers

14

Let's make one thing clear: That object on the front of the crankshaft is not a harmonic balancer—that term seems to be about as precise as say, panoramic brakes. Harmonic attenuator may be a more accurate description, but why not stop beating around the bush and call it what it is—a crankshaft torsional vibration damper, or for short, a crankshaft damper.

Torsional Vibrations

Torsional vibrations occur in a crankshaft due to the firing impulses delivered along the length of the crank, and the forces generated by inertia loads. Ignoring the crank's rotation for the moment, these forces generate movement of one big-end journal in relation to another. Although it may seem stiff, the crank is quite elastic and will twist up and then unload at a frequency dependent on its stiffness and the amount of mass involved.

Plucking a guitar string, or tapping a tuning fork will generate a certain vibration frequency; this we hear as sound. Sound is the natural resonant frequency of the system, and it de-

pends basically on two factors: the amount of mass involved, and the stiffness involved. Increasing the mass causes the natural resonant frequency to become lower; in other words, it oscillates more slowly. If the stiffness becomes greater, the natural resonant frequency increases.

At lower rpm a crankshaft is forced to vibrate simply from the input of the power strokes. The frequency that it vibrates at will, of course, be dependent on the frequency of the power strokes.

As rpm climbs, however, the frequency of the power strokes at some point will coincide with the natural resonant frequency of the crankshaft. When this happens the vibrations go out of sight because an impulse from one cylinder simply adds to the vibration of a past impulse. If left unchecked, these torsional vibrations can fatigue a crankshaft to death in short time.

How do we damp such vibrations? The basic answer to that is friction. A simple test to understand how friction affects vibration is to pluck a guitar string and then touch your finger on the string. The note will die because of the friction. Though you can't go around with your hand planted on the end of a crankshaft to kill vibrations, in principle, this is how all types of crankshaft dampers work.

Give a cast-iron crank and a steel crank a light tap with a brass mallet, and they'll emit a different note. The steel crank rings far better than the cast crank, the reason being that the material of a cast crank has far more molecular friction than steel. The net result is that cast-iron cranks are less able to support torsional vibrations than steel cranks, so that's definitely something in their favor.

It takes a lower amplitude of torsional vibrations to break a cast-iron crank, however, which tends to counteract this. Thus cast-iron cranks need damping just as much as steel cranks. When the limits of a cast-iron crank are approached, damper requirement becomes even more criti-

cal. On lower horsepower vehicles it's less critical, initially—almost to the point where any damper of any consequence would do.

Damper Types

There are three basic variations of crankshaft damper: the elastomer damper, the viscous damper and the pendulum damper. Although each is substantially different in design, they all rely on friction and relative motion between the crankshaft and the mass to do the damping.

The most common type of damper is the inertia ring elastomer-type damper. This type is fitted stock to all small-block Chevys when they leave the factory. It has an outer flywheel bonded by a rubber-like medium to the inner hub, which fits tightly onto the crankshaft.

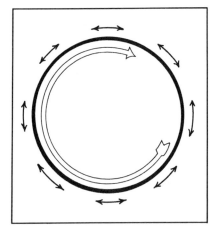

If you've never tried to envisage torsional vibrations before, then you'll need to use a little imagination here. Imagine you're looking at a hub on an un-damped crankshaft. Each firing impulse put into the crank is going to cause the crank to bend. The journal will first move ahead of where it's supposed to, and then the stored energy in it will cause it to spring back behind its neutral position. If we had a video camera that went around with the crank, we would see the end of the crank vibrating back and forth, as indicated by the small arrows. These vibrations, coupled with bending loads, ultimately lead to crank breakage.

The higher the engine's output, rpm or required endurance is, the more critical the crank damper becomes. Selecting a crank damper to suit the requirement not only means greater reliability, but also more horsepower.

Most rubber compounds have a high molecular friction, which means they won't sustain vibrations easily; they turn vibrational energy into heat due to their internal friction.

Elastomer dampers depend upon not only the mass of the outer ring, but also the elasticity of the compound. Small changes in the compound characteristics can significantly affect the damper's working range. Dampers that have aged and been subjected to high heat and heavy usage will have different characteristics than a brand-new damper. For this reason, you should regularly inspect the damper to see that it is in good condition.

Elastomer dampers can work well, but they do have one basic drawback—they are frequency sensitive. The elastomer material has a certain degree of springiness. Though it may be a poor spring, at certain frequencies this springiness can cause the damper to go into a torsional oscillation. The way around this is to design the mass of the inertia ring and select the elastomer characteristics such that the damper has the greatest resistance to vibration at the point in the rpm range where the crank has the greatest tendency to vibrate. In other words, the damper must be tuned to the crank and its associated parts, which we'll call the system frequency.

System Damping

At this point I need to emphasize the word system, because anytime you securely bolt something to the crank, you change the vibration frequency of the crank and the object you just bolted to it. Securely bolting something to the crank means that the mass that you're dealing with has changed. A crankshaft with a heavy flywheel will have a natural resonant frequency different from one with a light flywheel. In the same way, the weight of rods and pistons can have a similar, though much lesser, effect on the crankshaft vibration frequency. Also, if you've gone through and lightened your crankshaft, and had it heat treated, the natural resonant frequency of the crank will also have changed. So, if you go through your engine and modify this, that and the other, the vibration frequency of the finished product may differ considerably from that with which you started.

Since an elastomer damper is tuned to a particular frequency, it's possible that the damper no longer damps effectively at the critical frequency of the assembly. Fortunately, elastomer dampers have a reasonably wide range of operation, and it would be difficult to completely put the damping effect outside the range of even the most highly modified stock setup. However, the damping may not be nearly as effective as it

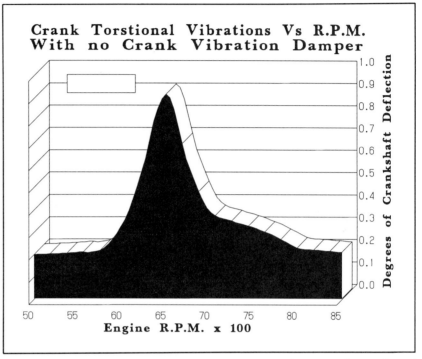

This graph shows what happens to crank vibrations as the engine runs through its rpm range. When the firing impulses coincide with the crankshaft's resonant frequency, the vibrations increase dramatically. The resonant frequency of this crankshaft assembly is about 6650rpm. At that point there was almost 1deg. of total deflection along the length of the crank. To put that into true perspective, the crank pins were flexing over 0.0625in back and forward from their neutral position.

Initial tests done to check stock dampers to find which is best showed the large dampers to be so. Subsequently, this was used as a base line to test the four dampers situated around it.

could be, which will shorten the life of the parts you so carefully prepared and invested in.

My crankshaft vibration damper tests involved first fitting this toothed disc to the damper.

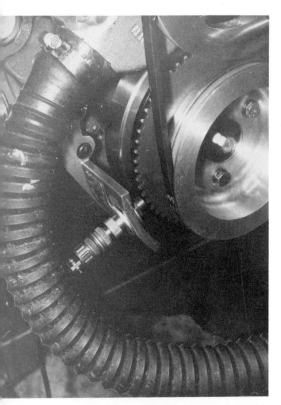

Next a magnetic pick-up was situated in line with the toothed wheel. The magnetic impulses from this were fed into this Furrior transformation analyzer. The resulting output was a plot of various vibrations throughout the rpm range.

Aluminum Hubs

When certain regulatory bodies outlawed the use of stock cast-iron dampers because of some damper explosions on high-rpm engines, many people simply opted for a plain crankshaft hub. In essence, this was a plain aluminum hub with no measurable damping capability. The argument was that being lighter with much less angular inertia, the hub would allow the car to accelerate quicker.

On the surface, the principle sounds good.

The viewpoint fails to take into account one vital aspect, however. If the front of the crank is experiencing the end result of severe torsional vibrations, such vibrations will be transmitted directly to the camshaft. Here's the glitch in the hub–increases–acceleration theory. Camshaft design has come a long way since the 1950s; a modern cam is the product of stringent mathematical equations which strive to produce the smoothest possible dynamics for the valvetrain, thus achieving the desired valve motion. All of these equations are based on the camshaft rotating at a constant speed or a constantly accelerating speed—not jerking around because of vibrations transmitted to it by the crankshaft.

Because of the 2:1 reduction from the crankshaft nose to the cam nose, assuming no damping takes place in the chain, we find that spurious vibrations are halved in amplitude. On a typical hopped-up street motor, at about 5000rpm a stock steel Chevy crank can flex as much as plus or minus ½deg., for a total of 1deg. deflection. When this gets to the camshaft, flex decreases to ½deg. of total deflection, and is superimposed upon the normal rotation of the cam. Flex plays havoc with the dynamics of the camshaft and virtually undoes all the careful design work that went into it.

Personally, I've found that if you use a high-quality camshaft from a reputable cam grinder, with advanced profile dynamics, the amount of power lost due to vibrations at the camshaft significantly outweighs the negative aspect of hanging the mass of a damper on the front of the crank.

On the other hand, if you use an inexpensive camshaft with indifferent dynamics, the effect of the crankshaft damper on power output appears minimal. Tests simulating a drag-strip run, and measuring the

power at the same time, indicate that selecting the right stock damper delivered superior performance on the drag strip as opposed to using a lightweight aluminum hub. This was proved, even though there was as much as 11lb difference between the damper and hub. Not only does the hub have a tendency to slow down the vehicle, but it also has a strong tendency to cause the crank to break far sooner than would otherwise be the case.

Damper

Over the years there has been a lot of controversy over what size of damper is optimum, especially for a drag-race engine. The thinking is that a large, heavy damper acts in the same way a flywheel does, and the extra mass in the damper means that the engine has to put more energy into the damper to be able to accelerate the additional mass. Therefore, there's less at the flywheel to accelerate the car.

The situation changes dramatically as engine rpm and horsepower start to escalate. I'm not talking about high-output race engines, but engines that are just hopped-up street units. If the engine's going to put out 280-300hp, and turn up to 6000rpm, then some attention needs to be paid to the damper. My tests indicate that the biggest and heaviest Chevrolet damper also proves to perform the best. The tests were held under accelerating conditions—not static conditions—and took into account the additional weight of the damper. The damper that proved superior, Chevrolet part number 3817173, was used on the 302 and 327 engines.

Since most people will be hopping-up a 350, there may be a perfectly good 350 damper on the engine already. And since these parts do cost money, if your original damper is good, then a 350 damper is sufficient, although it certainly is not the best.

During power testing it was found that the big damper actually made more horsepower under accelerating conditions than the aluminum hub—proof that you don't need an aluminum hub, however neat and colorful it may look.

To use a stock damper, check its condition. Unless it's brand new, you'll need to inspect it for any slippage of the outer ring, forward or backward, and that the elastomer is in good condition. Sometimes this is difficult to tell if the damper has been painted; the paint begins to crack. Get a wire brush and brush the paint off the elastomer; then check to see if it has any age or heat cracks. If it's less than perfect, it will do a less-than-perfect job.

If you're running a 400 motor, or a 383 utilizing a suitably modified 400 crank, then you have little option but to use the 400 damper as this damper has the requisite outer balance mass to externally balance the 400 crank.

If you're making damper swaps indiscriminately, another potential problem is that in 1968 the zero-degree mark on the crankshaft was moved. Basically it's about 15deg. off, which means that if you swap for a later damper and still use the same timing mark on the timing chain case, ignition time is going to be 15deg. off. You can get around this problem easily, though, by scrapping the timing tag welded to the cam-drive timing chain cover, and install a bolt-on tag. These are available to fit the 7 and 8in dampers, and are available at any good speed shop.

Aftermarket Dampers

In the mid 1980s a series of stock damper explosions prompted the outlawing of stock cast-iron crankshaft dampers at serious drag-racing events. It's questionable whether this was a wise move or not. At the time there were no directly recognizable alternatives, and this led to a rash of hubs being marketed. As I've said before, these hubs did not contribute anything to the damping characteristics of the crankshaft. As a result, there were a number of equally disastrous engine explosions due to crankshafts breaking.

The speed equipment industry was quick to respond to the need for an effective damper. Although a number of dampers were introduced to the market and tested, at the time only two dampers proved noteworthy. These were the friction or pendulum damper, designed by Doug Fischer, and the VibraTech viscous damper.

The beauty of these dampers is that they can be termed broad band devices. In other words, they will damp vibrations over a wide frequency, virtually eliminating the need to tune the damper to the system. Thus you could indiscriminately change the system stiffness by going to superior crank material, lightweight flywheel,

light rods and such. This would put the critical frequency up to considerably higher rpm. In spite of this, either of these two dampers would catch and damp the torsional vibrations developed.

The Fischer damper, which in my opinion is a well-made piece, works much like a clutch pack that's permanently set to slip. Inside the case of the damper, spring-loaded friction pads bear onto a surface directly connected to the crankshaft. If the motion of the crankshaft is smooth, then the friction pads accelerate as a unit with the outer case.

If vibration is present, then a back-and-forth oscillation will be superimposed on the rotational motion. This means the nose of the crank goes through minute, but rapid speed changes. Because of their inertia, the clutch-pack discs cannot follow this movement, so they slip, and the vibrational energy is absorbed and turned to heat by the friction generated.

The VibraTech damper, though completely different in design, utilizes a similar principle. But instead of friction between pads, it uses viscous shear, which is the internal friction of a liquid. Essentially, the VibraTech damper has an inertia ring in a case directly bolted to the crankshaft. When vibrations occur, the case will vibrate back and forth, but the inertia ring, being connected

only by the viscosity of the fluid, will not be able to follow the exact motion. As a result, a viscous shearing action occurs between the inertia ring and the case and this damps the vibrations.

VibraTech has been making dampers for diesels since the early 1900s, with literally millions of them in service. Over the years I've had a good deal of experience with the VibraTech damper, and I've done some of the original testing on their effectiveness for high-performance automotive applications, the small-block Chevy in particular. I've used these dampers on many subsequent test engines with great success. Vibration increased up to as much as 400 percent when the VibraTech damper was replaced with the best stock damper available.

In terms of damping ability, in theory, the Fischer damper should have as broad a scope as the VibraTech damper. Also, the potential for an elastomer damper to do the job, if tuned to the critical frequency of the crankshaft and assembly, should be equally as good.

The problem with an elastomer damper is making one that's precisely up to the job. To this end Callies Performance Products has an elastomer damper that is specifically designed to go with their crankshafts for stroke lengths up to 3.75in; for stroke lengths above that, they recom-

---

Damper Dyno Test

All tests run at 200rpm/sec. Acceleration rate.
All figures are an average of a minimum of four runs
Engine: 355 Small-block Chevrolet
Dyno: Superflow

RPM	Hub HP	Stock Damper HP	Viscous Damper HP
4500	351.1	360.5	360.7
4750	374.6	382.3	382.5
5000	401.1	408.1	407.0
5250	416.0	421.2	423.0
5500	419.2	425.7	424.1
5750	420.3	424.1	426.3
6000	421.3	425.6	429.6
6250	398.7	402.5	409.6
6500	369.2	375.7	379.6
6750	354.3	356.9	355.2

If you need concrete proof that hubs do not produce better results on the dragstrip, just read these figures. Even the stock damper on a simulated dragstrip run proved better than a lightweight hub,

in spite of the fact that it was some 11lb heavier. At the end of the day, though, the viscous damper beat the stock damper by quite a margin.

mend an alternative such as the Vi-braTech damper. The Callies damper is a good example of an elastomer unit able to do a good job because it's specifically tuned to the assembly. One of the advantages of designing both crank and damper in unison is that it allows the lightest possible damper to be developed, and a light-weight damper that does the job is going to have the same effect as shedding mass from the crank or flywheel.

Degree Marks

Virtually all aftermarket dampers have degree markings on them. Usually they are degreed up to 45-50deg. BTDC, and at the 90deg. points around the rest of the damper. For the most part, this is perfectly adequate for setting the ignition timing and valve lash. There are a couple of points worth noting, though.

Many timing lights are adjustable, with a dial that brings back the TDC mark to align with the pointer on the cam cover. The advance is then read off the timing light. These lights are not always as accurate as they should be, however, so it is better to use a fixed timing light to set the ignition timing on the damper.

If you've already established that TDC occurs at the TDC mark at the time of assembly, then you can be sure that your ignition timing is right when it's set to the relevant degree mark with the engine running. If the damper has full 360deg. markings, you'll find this a useful aid in setting and checking your cam timing.

If you're building an engine on a relatively tight budget, and you can't afford one of the aftermarket dampers, then adhesive timing tape is probably the next best substitute. Just how accurately timing tape indicates the angular displacement depends on the accuracy of the tape itself and the pulley OD. Though neither seems to be terribly accurate, the amount of error involved over say, the first 45-50deg. is relatively minimal. At least it's better than trusting the adjustable timing lights.

Damper Installation

The first rule of damper installation is to never get near it with a hammer. For a small-block Chevy, a B&B damper installation tool is a must, whether you're installing a stock damper or a high-tech aftermarket piece. Once you're equipped with the right tool, check that the damper is a tight fit on the crankshaft. This typically calls for an interference fit in the range of 0.0015–0.002in. If the damper—and this means *any* damper—does not fit tightly, it will not work.

The whole premise is based on the fact that the damper case must transmit the vibrations of the crankshaft to whatever inertia ring the damper has. In other words, the damper case needs to function as if it's welded on as far as the crankshaft is concerned. Besides ensuring a tight fit on the crank, use a suitable heavy-duty bolt to torque up the damper to 70lb-ft. Many early cranks prior to about 1968 do not have threaded snouts, so it may be necessary to drill and tap your crankshaft appropriately.

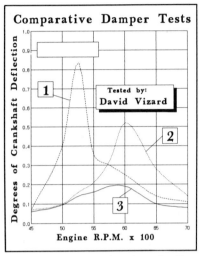

Comparative Damper Tests

Tested by:
David Vizard

When it comes to damping vibrations occurring at an unknown frequency due to modifications of cranks, rods, pistons, flywheel weight and so on, there's little that can beat a viscous damper. An elastomer damper can only match this type of performance when it is properly tuned to the vibration frequency of the system.

All crank dampers, but most importantly the viscous fluid dampers, should be installed with a proper installation tool, not a hammer. This item from B&B is highly recommended.

The B&B tool is used by screwing it into the end of the crankshaft, then tightening the large nut down onto the damper. This draws the damper onto the snout of the crank.

Wet-Sump Oiling Systems

15

The key to survival for any engine lies in the use of a quality motor oil. I've seen small-block Chevys used for day-to-day transportation that have run well over 200,000 miles, and still deliver quality performance. At teardown, some of these engines have shown so little wear that they could have run another 100,000 miles before their performance degraded to an unacceptable level—even for a performance enthusiast. What was the secret to their longevity? Certainly, many factors come into play, but the major factor has been regular oil changes at intervals as frequent as 3,000 miles, and the use of quality oils.

In terms of long life, for the small-block Chevy it's not the bearings or bores that normally present the problems. It's usually a case of the lifter or cam lobe wearing out—on many types of engines, the cam and lifter contact patch is the critical lubrication point of the engine. Bear this in mind when selecting oil for any small-block Chevy.

Oil Selection

There is no clear-cut answer to the question of what is the best oil. Some oils have superior attributes to other oils; one oil may win out in one area, and another oil in some other area. As far as deciding which is the best oil, the choice boils down to what suits your particular application. For what it's worth, here are some guidelines based on experience that I feel important to qualify.

First, most of these engines live and die on the dyno—that's what they are built for. They're used as mule motors until such time when either they suffer a major catastrophe due to too much rpm, or I feel a teardown is necessary to restore them to the condition of a dyno motor competent to do consistent back-to-back tests. This means that the oil in these engines is never called upon to do much in terms of hours of running. It is expected to take a beating as far as loads, rpm and temperature are concerned.

After trying several oils I settled on Valvoline because no major oil-related problems have surfaced to date, and I have ready access to it at a local supplier. Consequently, top-grade Valvoline oil goes into most of my dyno engines, unless I'm specifically testing lubricants.

As for engines used out on the track, I've used Castrol for many years with great success, not just in small-block Chevy engines, but in a variety of engines, including turbos.

I've also used synthetic oils. The so-called advantage of synthetic oils, such as Mobil One, Cam Two, Amsoil and others, is that they're supposedly more stable in terms of temperature-related viscosity. They don't decompose as easily or oxidize at high temperatures like conventional mineral oils do, and their lubricity is supposedly greater; in other words, they're slipperier. I've had experience with Mobil One and Amsoil in some race engines and have had no problems.

I have heard, however, of instances where a greater rate of cam follower failure occurred with synthetic oils than with conventional mineral oils. This seems to apply only to motors with flat-tappet cams, which is not an unexpected situation. As pointed out previously, the highest unit loading in the engine is probably between the face of the cam lobe and the face of a flat follower. Roller follower engines would not suffer from this problem.

I'm inclined to believe that the reason I've not seen an increase in the cam-follower failure rate is that when the engines are built, I almost always use Crane Cam Break-In Lube. This contains zinc-dythiophosphate, which is an excellent high-pressure lubricant. It is an additive commonly used in many mineral-based oils to boost the oils' high-pressure lubricating abilities. Although it may prevent scuffing and rapid wear of cam followers and cam lobes, if it is used in excess it can cause cam follower pitting. This should not be a problem with an engine that's only going to run 500

miles, but if you're building a street motor and you expect to go 50,000 miles plus between teardowns, do not use too much of this additive in the oil.

A teaspoon or two of the Crane Cam Break-In Lube at each oil change is probably all that is needed. If cam follower problems still exist, then it's more than likely that you're running too much spring pressure, or a follower is out of alignment with the cam or some other similar mechanical problem exists.

I have used GM's EOS engine oil supplement and results suggest it's one of the few additives worth recommending. However, make sure your engine is oil tight in the rings and valve stem seals department. EOS contains compounds that, if allowed into the combustion chamber can,

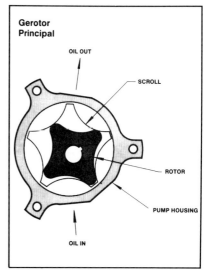

This cutaway of the gerotor pump reveals its inner workings. Both the rotor and the scroll rotate, but they're on different centers. The oil is drawn in at the end of the case in the lower position indicated, and fills the cavity between the rotor and the scroll, which progressively gets larger until it reaches about 180deg. to the other side of the entry point. At this stage, the cavity starts to get smaller again, due to the eccentricity of the two components, and the oil is pushed out of the cavity.

after being burned, glow and cause pre-ignition.

Power Versus Oil Type

One of the big questions that comes up anytime lubricants are discussed is whether or not one oil will produce more horsepower than another or indeed, whether pouring an additive into the engine will help horsepower. It's impossible to make a blanket statement here, since so many variables present themselves, and there are so many products on the market that I have not yet tested. Although I cannot be specific, I can pass on my experiences.

The synthetic oil manufacturers claim that synthetic oils will produce more horsepower because they are slipperier. This may or may not be the case. I have not tested whether they are, in fact, slipperier or not, but one thing I have found evident from my dyno testing is that the thinner the oil, the less its viscosity, the more horsepower the engine will create. For instance, running a mineral oil at 200deg. F. and then increasing the oil temperature to 220deg. produces as much as 3-4hp gain in a 350hp engine.

The problem is that running mineral oil at 220deg. is pushing it near its temperature limits. Certainly, going as high as 250deg. F. is treading on thin ice; however, synthetic oils seem to be quite comfortable up to temperatures in excess of 300deg. Apart from this, their viscosity does not change as much with rising temperature as does that of mineral oils. This means that you can use an oil that is thinner to start with, and get the benefits in terms of extra power that a less viscous oil will deliver.

As near as these tests can determine—and here we're looking at the limited accuracy of the dyno—I could see no measurable difference in synthetic oils and mineral oils when both were run at similar temperature-developed viscosities. I felt, however, that the mineral oils used were often a little low on viscosity for normal use. If a mineral oil was run at a suitable temperature to bring it to the viscosity of the synthetic oil, then I found that the increased temperature required to reach such a viscosity caused a drop in power.

Many of the test results indicated that with mineral oils, coolant tem-

perature needed to be 170deg. F. and lubricant temperature around 220deg. It is difficult in practice to get the oil temperature this much hotter than the coolant temperature. When synthetic oils were used I found that the engine would deliver the same kind of horsepower with lower oil temperatures. The sort of temperatures seen were more practical for use on the street or track.

Now the question is, how much potential exists for additional output when synthetics are used? An average test engine is usually around 400-450hp, with an rpm limit around 6500-7000. Under these conditions synthetic oils can be run 20-40deg. cooler, and between 2-4hp more seems to result. Such numbers represent small power changes, especially in terms of engines that produce around 400hp and in reality, they represent the limit to test accuracy. The figures are the result of averaging out a number of test runs. Rather than be specific and quote power figures, it's probably wiser to say that, Yes, synthetic oils are worth horsepower, this additional power being on the order of 0.5-1 percent.

There is a side issue here, however. With the ability to run the oil cooler, it's possible to cool piston crowns better. This could mean that slightly higher compression ratios could be used, and pistons could be made slightly lighter because the temperature of the crown will be lower by virtue of the cooler oil splashing the underside.

Synthetic Oils

One of the problems with running synthetic oils is the initial expense. There are several types, and they can be differentiated by the base from which they are built. The most common synthetic oils for the automotive industry are those built from factions normally found in crude oil. By built I mean that the molecules of various components are broken down and then rebuilt, usually into longer chain molecules designed to have a specific characteristic.

Some oils which, for want of a better word, are even more artificial, are built from polyesters and diesters. These oils are the most expensive, and are usually used in the aircraft industry for specific heavy-duty applications.

For the most part, racers will be looking to use the mineral-oil-base

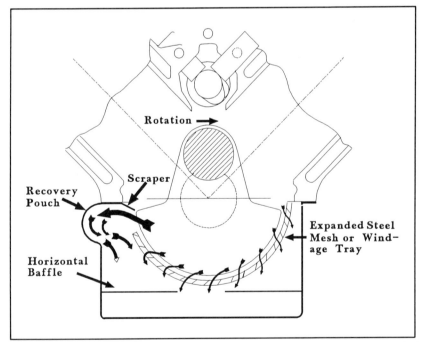

Here are the basic elements of a high-efficiency wet-sump system. The sort of pan depths shown here are what would normally be used for a circle-track machine. Much of the additional oil could be held in side pouches, not shown here. The most important aspect to appreciate is that the scraper and the expanded steel mesh pull the oil from the crank and return it to the pan in such a way that it won't get tangled up in the crank on the return to the pan. For drag racing, the pan would be much deeper than is shown here.

synthetic oils which often can be diluted as much as 50:50 with regular mineral oils and still retain better than 50 percent of the advantages of the synthetic oil.

In other words, a 50:50 synthetic mineral oil blend's performance will often come closer to that of the pure synthetic than of the pure mineral oil. Of course, this will vary from oil to oil. Some tests conducted with a French oil, Motul, which is generally not available in the United States, proved effective in this area. A 50:50 dilution significantly cut the costs of the oil, but did not significantly reduce its effectiveness. Of course, diluting the oil is always going to make it less capable, so if you can afford to use straight synthetics, go ahead and do so. But there are a few facts you should consider.

First, synthetics seem to work better with roller cams. As we've already discussed, evidence indicates they don't seem to be able to hold up on the high-pressure lubrication side, required at the cam lobe to follower interface—at least as far as small-block Chevrolets are concerned. This may be a problem peculiar only to the small-block Chevrolet because the same problem with flat-tappet cams has not been witnessed in many other types of race engines. It could be due to lack of sufficient testing in that area, or to just dumb luck.

When changing to synthetics it is necessary to allow for their considerably thinner consistency under cold conditions, as opposed to mineral oils. Although the viscosity doesn't change much, a suitable grade synthetic may still be thinner than a regular mineral oil at operating temperatures. For this reason I recommend that race engine bearing clearances be marginally tightened up; I'm talking about 0.0003-0.0005in. If you're going to run a mineral-synthetic oil split, then maybe the original bearing clearances that could normally be run with a straight mineral oil should be kept.

I emphasize here that this is not a hard-and-fast rule, and in the light of further experience and information I may change my opinion. Current practice does seem to follow what has been suggested here, however, and it seems to be working.

Oil Temperature

It's taken for granted that oil provides lubrication for the engine, but another primary function of oil is that it's a cooling medium. The pistons especially rely on a significant amount of cooling from the oil splash on the underside. Although we try to keep as much oil as possible away from the whirling parts of the bottom end, a certain amount has to be present for cooling purposes.

This leaves us with the inevitable tradeoff situation: If the oil is allowed to get too hot, it may be better initially for reduced lower-end losses, but also it can lead to overheating problems. Even if no oil film breakdown occurs at the bearings, pistons can weaken due to the elevated temperatures they're running at, plus the engine is more likely to detonate. On the other hand, cooling the oil too much means that there's going to be too much power lost to viscous drag.

Generally, the higher the rpm concerned, the more piston heat we have to deal with; therefore, more oil is probably needed on the underside to cool the piston. But by the same token, there is a greater necessity to keep oil away from the crank at these higher rotating speeds simply because the losses go up with the square of the engine's rpm. So, doubling the rpm means four times the windage loss. In other words, if we can shave off 2lb-ft of windage loss at a 4000rpm, this will inevitably represent about 8lb-ft at 8000rpm. That's 1.5hp saved at 4000rpm by making some improvement in the bottom end, but this escalates to a whopping 12.2hp at 8000rpm. What is the best tradeoff likely to be, then?

In a normally aspirated engine, ordinarily there is enough oil going around underneath the pistons to be able to adequately cool them, especially if a synethtic oil is used and bulk temperatures are kept down to the 170-190deg. F. mark by means of an oil cooler. If temperatures are allowed to escalate much over that, there may be a penalty to pay in terms of having to run a thicker crown piston, slightly lower compression and so on. If the subject is a supercharged or turbo motor, then an entirely different situation presents itself.

Cooling Supercharged and Turbo Engines

With supercharged engines, the high rpm usually seen by normally aspirated engines is rarely used. Instead of rpm, boost pressure is used to generate the horsepower. Of course, rpm is still an important factor, but supercharged small-blocks inevitably run quite a few rpm less than their normally aspirated counterparts. Thus, the losses from the windage is less and, to a certain extent, cooling the piston becomes more important than keeping oil away from the piston-crank assembly to cut windage. So it's reasonable to assume that if the intention is to build a turbo or supercharged small-block Chevy, then provision for oil cooling on the underside of the pistons can pay power dividends.

The easiest way to do this is to drill some auxiliary oiling holes in the upper part of the main-bearing housing in the block. This involves drilling through the oiling groove and aiming the hole at the underside of the piston. If the oil hole is directed toward the centerline of the piston, just in front of the connecting-rod small end when the piston is halfway down its stroke, then, as the piston moves up and down the bore, the spray of oil will impinge across the width of the piston. From here it usually runs across the underside of the piston crown and back down into the crankcase. This effectively draws a substantial amount of heat out of the piston crown.

Although I've not done any back-to-back instrumented tests, I would hazard a guess from hardness-checking pistons that crown temperatures are reduced somewhere between 40-100deg. F. which, incidentally, on supercharged engines are much higher starting out than on a normally aspirated engine. Using the oil to cool the pistons, temperatures can be lowered to those normally seen by unblown engines.

Another way to cool piston undersides is to use rifle-drilled rods such as those made by Cunningham Rods in Gardena, California. Though I've not tried this personally these rods, normally drilled to oil the pin from the big end, could be easily modified to spray oil from the top of the rod. To make up for the added oil-flow requirement, use a higher volume oil pump.

Oil Pumps

The stock small-block Chevy pump as it comes from the factory proves to be both reliable and satisfactory; however, that does not mean it's perfect. Some racers seem to feel a desperate need to install a high-pressure, high-volume pump. In many cases this is simply not necessary.

Figure the stock oil pump is good for power outputs up to 350-375hp, and rpm of 6500-6750. Therefore, it's not necessary to do anything to it unless you want to improve its overall performance. A stock oil pump will function adequately and keep the bearings in your engine alive—about 50lb of oil pressure is all that's required.

It is possible to have too much oil pressure, though. All this achieves is higher oil temperatures at the expense of additional power to drive the pump. You can look at it this way: the higher the oil pressure is, the more back pressure there is on the pump trying to stop it turn, therefore the more power it takes to turn it. Not only is excess pressure an obstacle to power production, but also excess volume.

If the stock pump has adequate volume, installing a high-volume pump, which probably bypasses all the additional oil volume, serves only to *reduce* power output. Consider whether or not your application needs anything other than the stock pump. If it doesn't, you won't be improving the situation by spending your money on a higher volume or a higher pressure pump.

Although the stock pump is more than adequate, it can still be improved upon. If you rework the oiling system, or even just the main cap where the oil pump bolts, you will have made some improvement to the way the oil is discharged into the block. Improving the oil flow into the block means there would be less pressure lost between the pump and the bearings, and after all, it's not so much the pressure in the galleries that we need to consider as the pressure at the point of entry into the bearings.

Internal Pump Clearance

If you have several pumps to choose from you could possibly put together one with tighter clearances. For instance, the normal clearance between the edge of the gear and the outer case is between 0.002-0.004in. By selecting the housing and appropriate gears, you could put a pump together which has 0.002in clearance instead of 0.004in. This will reduce the leakage around the outside face of the gears during the pumping action. Also, the height of the gears in this case can be important. Again there is a tolerance between 0.002–0.004 in.

If the gear faces are 0.004in below the face of the pump, then the face can be rubbed down until the clearance is reduced to 0.002in. Again this will cut the amount of oil that's spilled from between the gears as the pumping action is delivered.

If bearing clearances are increased in an effort to get the crankshaft to run freer, then tightening up the pump clearances can be beneficial as far as the idle oil pressure is concerned.

The most difficult oil pressure situation to satisfy is at low, not high rpm. It's often a misconception that engines need more oil as the rpm increases; in fact, the oil pressure requirement is not necessarily in proportion to the rpm increase.

The clearances do not significantly increase as rpm goes up, but the pump's output does increase. Thus, the oil pressure is likely to increase as rpm climbs unless it is limited by the pressure relief valve. This could mean that a pump that is inadequate at 4000rpm is more than adequate at 8000. However, the engine has to pass through the 4000rpm mark to get to the 8000, so the pump must be sized for the worst conditions—that is, satisfying enough oil pressure for low-rpm usage. Consequently, all oil pumps are oversized for the top-end rpm of the engine, and because they have to be sized for the low-rpm applications, the oil bypass system assumes a relatively important role.

Oil Pump Chatter

There are many different brands of replacement oil pumps on the mar-

What you're looking at here is the discharge hole from the pump, and the fact that the gear teeth tend to shroud it. The discharge hole needs to be modified as described in the text.

This Moroso pump has been modified by the inclusion of anti-chatter grooves. It's based on the heavy-duty Melling pump.

ket, apart from those that are put out by Chevrolet. Although they may look similar, there are differences you should be aware of. One of the major characteristics of a small-block Chevy pump—one that we need to try and minimize—is a pulsating oil flow which can cause what is known as pump chatter.

Anytime the oil pump doesn't rotate smoothly, a back-and-forth vibration is superimposed upon its rotation. This same vibrating motion also hits the distributor since its drive originates from the same point. In turn, this can cause spark scatter and inevitably leads to a reduction in power output.

There are many so-called cures for spark scatter. One remedy simply involves using a die grinder or a high-speed electric drill. If you remove the top cap from your pump and inspect the outlet hole down in the cavity, you will see that on some pumps the driving gear can cover part of the hole. As the driving gear rotates, a tooth will begin to cover up part of the hole and then almost completely uncover it before the next tooth starts to cover it again. This causes the outlet area to change as the gear teeth go by it. As the gears go around, the area for the oil to flow out of the original hole can vary as much as 18 percent.

Not all pumps operate in this manner, however. Some pumps have the gears positioned so that they are slightly above the hole, creating a depression all the way around the periphery of the gears on the floor of the pump case. This tends to minimize any adverse effect on flow through the discharge port. If the gears on your pump intermittently cover the hole, then the best plan is to grind a nice lead-in to this discharge port with a bias directed toward the idler gear.

Irrespective of whether or not the gear covers the discharge hole, it does pay to streamline the entry into it because flow from the point where the gears compress the oil between the teeth is very disoriented. Cleaning up the discharge outlet will not only help the flow out of the pump, but it will also cut the pressure loss at that point, therefore making it slightly easier to drive the pump. The effect is minimal, but for what little effort it takes, you might as well do it.

A popular way to reduce the chatter on a stock seven-tooth small-block Chevy gear pump is to cut antichatter grooves into various parts of the body. How these grooves operate to overcome chatter has never been satisfactorily explained, so for want of a better theory, I'll simply give you mine.

Essentially, it seems that the antichatter grooves operate on two principles. First, they apply pressure to the end of the gears which, in turn, press them onto the end of the case and add friction, tending to damp out the chatter. Second, the grooves spill off some of the oil as it's compressed between the teeth, thus making the trapping pulse as the teeth mesh together less severe. If this is the case, then any antichatter grooves may have the effect of reducing the pump's output. This, of course, won't be of any consequence if the pump's output already exceeds requirements.

Pressure Relief Valve

Most stock small-block Chevy pumps deliver an oil pressure of around 35-45psi. On occasion, I've found that changing to synthetic oil on an engine having the pressure relief valve set at a fairly low pressure—around the 35psi mark, for example—will cause the oil pressure to drop as much as another 5psi.

If you're rebuilding a pump with a known history of low oil pressure and you're certain it's not a bearing clearance problem, and if the innards of the pump look to be in good conditon, then you should suspect a weak pressure relief spring. The simple answer is to buy a new pressure relief valve spring and compare the stiffness and length to the existing one.

If you want to increase the oil pressure, this can be done either by fitting a stronger spring, Chevrolet part number 3848911, or you can simply shim up the spring that's there. If you elect to shim the spring, be careful that you do not limit the travel of the piston such that it will not bypass the oil adequately.

Oil Bypass

On a stock small-block Chevy pump and most replacement wet-sump pumps, the oil is bypassed internally. This means that if the oil pressure at the outlet side goes above the setting of the bypass valve, the bypass valve opens, and returns the oil to the intake side of the pump. So when the engine is operating under a high bypass condition, which it does at high rpm, the same oil may circulate a number of times before actually being used by the engine. This causes the oil to heat up unnecessarily.

Though the amount of heat put into the oil does not constitute a major problem, steps can be taken to reduce it and increase pump efficiency at the same time. Within reason, the larger the return to the inlet side of the pump, the better. On most pumps this can be improved by taking out the press-in plug on the cross-drilling that connects the pressure relief valve drilling with the intake. You can also help improve the return flow by accessing this area with a die grinder and radiusing off the edges of the cross-drilled hole.

There is an alternative to dumping the bypassed oil back to the inlet of the pump, which for some reason or another doesn't seem to be done often. Rather than bypass the oil back to the pump and cause unnecessary heating, the oil can be bypassed directly back into the pan. Rarely do I see stock pumps modified to do this, although there's nothing complex involved. Pumps that are built to bypass the oil straight back to the pan are marketed by Faria Engineering, System One and Callies Performance Products.

If you have access to a few simple pieces of equipment, it's not difficult to modify an existing pump to bypass directly back to the pan. Essentially, this involves blocking off the original transfer hole between the pressure

The external bypass of the System One pump is seen here with the screen removed. The pump also has an adjustable pressure relief valve. This is adjusted by means of the set screw and lock nut at center right.

133

relief valve and the pickup pipe drilling. When plugging the hole, be sure that the fit is totally air and oil tight. In an air leak occurs, it will reduce the effectiveness with which the pump draws oil. If the air leak is large enough, the pump will not draw any oil at all.

Once the hole is blocked off, dump the excess oil back into the pan. Here you need to consider how the oil enters the pan. When the pressure relief valve opens it can send a high-

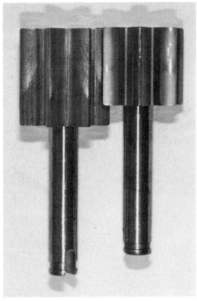

High-output small-block pumps employ gears 0.250in longer. Here's the difference between stock and high output.

pressure stream of oil straight into the pan. This can splash the pan oil considerably, thus I recommend putting the oil through some kind of grid to remove some of its kinetic energy so that it drops into the pan and creates less splashing.

Another method would be to couple up a pipe to the pressure relief valve passage and direct the oil to impinge on the side of the pan. Adopting this technique achieves two things: it cuts the temperature of the oil, and it appears to give a marked reduction in pump chatter.

This indicates that one of the possible reasons for pump chatter is that the bypassed oil fed through the transfer passage is pulsing and feeding that oil back into the intake side, which is also prone to pulse at the same frequency since the same gears are causing it. The result is too much oil supply pulsing, which could be a contributing factor toward generating unwanted gear chatter.

By externally bypassing the oil you can create a problem on the induction side of the pump as flow through the pickup pipe is greatly increased. This can cause the pump to cavitate and momentarily starve the bearings. To combat this, the stock ½in pickup, which at best is marginal, needs to be changed for one of ⅝in diameter.

Alternative Oil Pumps

Though there are ways and means of uprating a stock pump, there are also good reasons for utilizing a

pump with different characteristics. For instance, if bearing clearances have been increased, and the engine is using a low-viscosity oil or high operating temperatures are expected, then the situation may warrant the use of a high-pressure, high-volume pump. We've already discussed the technique for increasing the pressure of the stock pump, which may be a wise move if high rpm is to be used and bearing clearances are to remain relatively stock. Under these conditions 55-60psi of oil pressure is desirable, but certainly no more than this.

One of the major manufacturers of oil pumps is Melling, whose products are widely available throughout the country. Melling manufactures a stock replacement pump plus a high-volume pump, part number 55HV, which delivers 27 percent more volume. This pump uses extended seven-tooth gears and is approximately ¼in longer than stock.

Melling pumps are also equipped with stronger springs to increase the point at which the pressure relief valve comes in. Springs available for the Melling pumps are color coded according to stiffness. They come in plain colored, yellow or pink, and their stiffness increases in that order. The plain-colored spring sets the oil pressure to around 40-45psi; the yellow one to about 55; and the pink one to 65-70psi. Of course, these are hot oil pressures. When the oil is cold, pressures will be considerably higher.

If you're installing a replacement high-volume pump, be aware that these pumps are longer. If you use the stock pickup, it runs into the bottom of the stock pan so an appropriate Melling pickup must be used.

Another alternative is to use a big-block pump for a regular small-block Chevy. The big-block pump has about 30 percent greater flow rate than the small-block pump, and it has a twelve-tooth gear arrangement instead of the normal seven-tooth gear of the stock pump. Suitably prepped with an external bypass, this pump can reduce spark scatter substantially.

If you're going to go to the extent of buying a big-block pump and then make considerable modifications and if you've got the extra cash, you may want to save yourself some time and trouble and go straight for one of the external bypass pumps marketed by Faria Engineering, System One or Callies. These pumps employ twelve-

Here are the caps of two pumps. On the left is the latest Melling design as of 1991, and on the right an earlier design typical of most stock replacement Chevy pumps. The Melling one has a larger bypass hole and a sink around the bypass return in an effort to cut gear chatter.

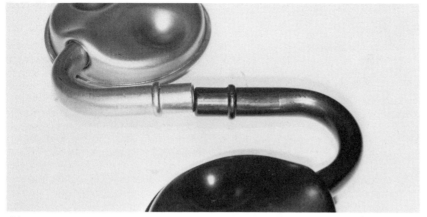

A longer pump means a different pick-up if you're using the stock pan, as there's approximately 0.250in difference in installed height.

tooth gears as per the big-block Chevy pump, and apart from bypassing externally, they have adjustable pressure regulators built in. If a suitable pipe plug was installed into the side of the pan, by simply draining the oil down to an appropriate level and removing the plug, the oil pressure could be adjusted up or down to suit the situation. Of course this would entail making a special tool, but that's not beyond the ability of any self-respecting racer.

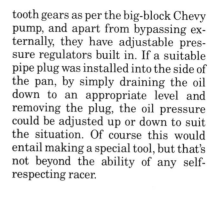

The advantage of the big-block pump is that it has more gear teeth than the small-block, and runs much smoother, thus cutting gear chatter.

Here are the top and bottom views of a big-block and small-block pump. Needless to say, the left pump is the big-block one. This pump can be adapted to the small-block using the Melling big-block Chevy pump with its relevant small-block pick-up.

Here's a cutaway of a high-output, bypass to pan, fully roller pump from System One. The aluminum body is machined on a CNC Mill, as is the cover, which contains a pressure relief valve adjusting screw.

For reasons we'll go into later, you may be contemplating using a deeper sump on your engine. This means the oil pickup point will need to be moved down to reach the oil. The pickup should be within ¼ to ⅜in at the bottom of the pan.

There are two ways to achieve this: extend the pickup pipe, or extend the pump. Extending the pickup pipe seems to be the easiest way; however, pumps work better on the pressure side than they do on the suction side, so it's not a good idea to put a long pipe on a pump mounted in the stock position. Longer pickups are sold by Moroso Performance Products and others in the sump pump business. Though an extended pickup is functional, it should always be considered the number two choice.

A far better solution is to lower the entire pump. This can be achieved with pump extension kits, and usually companies such as Moroso that manufacture specialty pans also market appropriate pump extension kits.

Sump Design

When the small-block Chevy was originally designed, the sump was simply a place to store the oil. But as performance levels of both chassis and engines increased, greater demands were made for efficient operation of the sump.

Principally, there are two areas where sump design can fall short. A car that's capable of accelerating or cornering rapidly can cause all the oil to move up to one end of the pan and away from the pickup. The result is loss of oil pressure and shortly thereafter, loss of the engine. So if a wet sump is to be retained, make sure the sump will provide the requisite oil supply to the pickup under all conceivable conditions that the car is likely to be subjected to.

In addition, it is a good idea for the sump to have some control over the windage oil. It is easy to mistakenly assume that since a crankshaft is whirling around at high speed, the oil gets centrifuged off it.

In fact, this does not happen, as explained earlier. Bill "Grumpy" Jenkins best described this action when he likened the oil form to ropes. Sure, oil does get thrown off by the rotating mechanism, but much of the oil hits the walls of the crankcase or pan and bounces back into the rotating parts.

Oil Pan Modifications

For a stock or near-stock motor, crank windage and oil entrainment in the crank may not be a major concern.

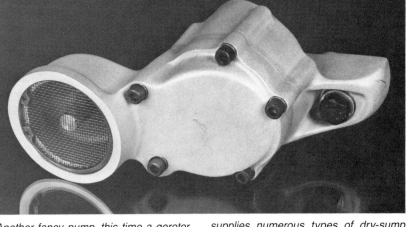

Another fancy pump, this time a gerotor design, is made by Barnes, who also supplies numerous types of dry-sump pumps.

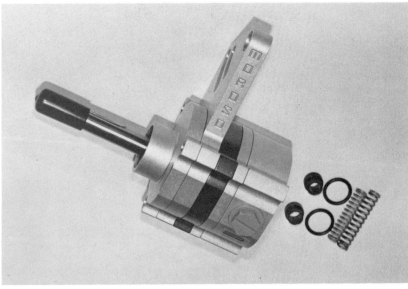

Exotica for a wet-sump setup is this external gerotor pump from Moroso.

The Moroso pump is expensive, but if you take a look at the internals you can understand why.

But let me put things into perspective. Tests were run on a 400ci engine with a calibrated sight level in the drain plug hole. Oil was progressively added to the engine between power tests. At a little over ½qt. above the full mark, the crank must have just started to dip into the oil. Normal windage in the pan and dispersion of the oil through the rest of the engine usually caused the oil level to drop at the sight plug by about ¾qt.

When the engine was a little over-filled, however, 4000rpm proved a critical speed where oil entrainment by the crankshaft suddenly became significant. At this point, the sight plug oil level dropped by almost 2qt., and power decreased by nearly 10hp. When you consider that most street motors regularly go past 4000rpm, you can appreciate that losing 10hp is significant.

Now you may say, "Well, I'll make sure my pan never becomes over-filled." But if the pan is not baffled, the fact that the car has accelerated may move the oil to the shallow part of the pan. There it can be picked up by the crankshaft, inducing the same situation as seen on the dyno, irrespective of whether or not the oil pan is overfilled.

The 400ci is not the only engine that has exhibited a power drop traceable to oil entrainment. These days, with modern tires capable of delivering good traction, and the fact that virtually any small-block Chevy can turn over 6000rpm, the oil pan needs attention, even for the most minimal performance application.

Just because only one part of the crankshaft starts to dip into the oil does not mean that entrainment stays local to that area. For instance, accelerating hard away from a stop light, the oil level may rise at the back of the pan and the rear counter-weight may start gathering oil. The most obvious conclusion is that oil will be entrained only at the rear of the crank.

In reality, because of the position of the counterweights, the crankshaft acts as a crude propeller and drives the oil forward. The crank starts to form its characteristic rope-like tentacles which extend along the length of the crankshaft. So even though the crank may have only touched the oil in one place, the windage problem and viscous shear of driving the crankshaft through all this oil, occurs down the entire length of the crank.

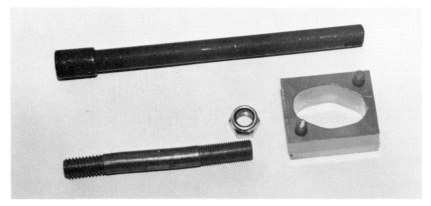

If you're using a deep pan it is better to extend the pump into the pan than to simply extend the pick-up. This simple kit from Moroso allows just that.

Here's an external view of the full-length recovery pouch on a 350 circle-track motor.

This is a view looking into a deep Pro Cam pan. Note how horizontal baffles are situated around the pump aperture.

For street vehicles the most obvious solution is to put a horizontal baffle on the pan in an attempt to keep the oil away from the crankshaft. If ground clearance problems don't exist, then use a deeper pan; the farther away the oil reservoir is from the crankshaft, the better. If a deep enough pan can be installed, then use a sheet-metal baffle with appropriate drain-back holes to separate the oil from the spinning crank.

If a sheet-metal windage tray is used, make sure that excess oil is scrapped from the crank. Scrapper blades should be placed along the side of the block where the crank is coming up to meet it so as to shear off excess oil. Direct the excess oil underneath the windage tray into the reservoir area below.

If the pan is shallow and there's little room for effective scrappers, then a slightly different approach is necessary. If a solid windage tray is used, the oil can hit the windage tray and bounce back into the crankshaft. On the other hand, if a wire screen is used, it absorbs the energy of the oil and lets it pass through the screen with far less tendency to bounce back into the crankshaft. Shallow pans then can benefit from the use of a mesh screen, whereas deep pan motors and those with effective scrappers and/or recovery areas work best with sheet-metal screens.

In an effort to reduce the amount of oil that remains on the walls, screens and baffles, some companies Teflon-coat the baffles and pan intervals. This encourages the oil to drain back into the reservoir area quicker.

Making an effective sump is quite a science, and although making your own is not beyond the skills of anyone handy with a welding torch and sheet-metal cutters, guaranteeing top results is! I have successfully used pans from Canton, Moroso and Ray Baker's Pro Cam series. These companies, among others, market a wide range of pans for various applications. Check their catalogs for one that suits your particular requirements.

Oil Filters

The original 265ci small-block Chevrolet was introduced without an oil filter. The following year the canister-type oil filter setup was added, and was used until 1968. From 1968 on, all subsequent small-blocks used a different adapter in the filter housing and a spin-on-type filter.

For most purposes, the earlier filter was significantly better; depending

Apart from the main scraper, the Pro Cam pan also has multiple scrapers mounted on adjustable threaded pillars.

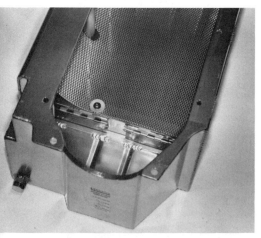

Oil into the pick-up area on the Moroso deep sump is prevented from backflowing out to the rest of the pan by the one-way doors seen here.

This Moroso circle-track sump pan follows the same pattern and design techniques described in the text.

138

upon the element used, condition results varied, but in general the flow of the elements used in the early filters was higher. In practice these early filters provided a high degree of filtration: 90-95 percent of the oil that went through the bearings also went through the filter.

The later spin-on cartridge filters eased filter changes, but they have a lower capacity and cannot handle the full output of the pump, especially when the oil is thick during a cold-start situation. To accommodate this an oil bypass valve is located in the filter spin-on housing bolted to the block.

The bypass valve usually lifts off its seat at a maximum of 14-15psi, but when the oil is cold it is not uncommon for the pressure differential across the filter to exceed the bypass release pressure. This results in cold, unfiltered oil going direct to the bearings. Since the oil pickup is at the bottom of the pan where most of the debris settles out after shutdown, we find that the filter is bypassing oil at a time when its filtering action is most needed.

At this point you may be tempted to block the bypass valve. That's easy enough to do. You can either make

the valve solid by plugging the bypass valve completely, or you can buy a kit from many aftermarket manufacturers that replaces the entire adapter with one having no bypass valve. Before you're tempted to do this, consider all the pros and cons, and the possible problems you could cause.

First, plugging the valve and using a stock size and flow filter can mean a bigger pressure drop across the filter. This can reduce the pressure fed to the bearings. Also, the bypass valve is likely to come into operation sooner so the quantity of oil available at the bearings is likely to be reduced as well. In addition, if the engine has a high-pressure pump, the pressure generated is likely to burst the filter, so now your oil's being pumped out onto the street rather than through the bearings. If the filter doesn't burst, then consider that the extra resistance to flow through the filter means more stress on the pump and usually more power to drive it, minimal though it may be.

What's probably a better alternative, especially if you're working within a budget, is to use a large small-block Chevy truck filter. With this you can expect to achieve the filtration performance typical of the early canister filter.

If you have a later-style block you can get an adapter kit, Chevrolet part

number 5574538, which allows the use of the early canister-type filter. Otherwise, you can use an aftermarket filter. If you want to stick with a more or less conventional paper-type filter, then K&N Engineering offers one of the best on the market. The case is able to withstand 450psi, it has a hex underneath on the lower face for fitting and removing, plus wire tags for wiring the filter on to ensure that it won't shake loose. The filtering element has high-flow and high-filtration capabilities, which make it ideal for racers. It has good flow capabilities and flows about 2½ times more than a typical OE filter while taking out particles up to about 40 percent smaller.

The Fram HP filter is also a good one with about twice the flow of a normal paper filter along with the K&N excellent filtering capability. Sure, it costs a little more than a regular filter, but you can't get additional quality without additional expense.

If you're prepared to shell out a little more cash yet, there's another type of filter available—the stainless-steel screened mesh filter. Oberg and System One both manufacture stainless screened mesh filters. The Oberg filter consists of a flat screen sandwiched between two plates and is meant for remote mounting. The System One filter can be used as a direct

I've not tried all brands of oil filter, but my experience indicates that the K&N is amongst the best available, being able to outflow most stock OE-type quality filters by at least 100 percent.

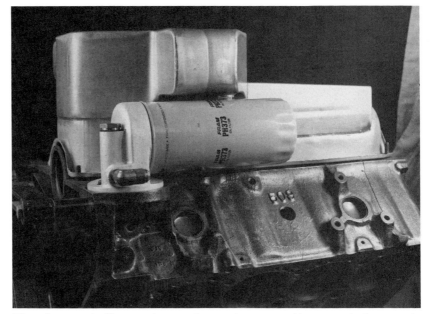

When it comes to filters, so long as the filter media is good, then bigger means better. This long Fram filter avoids ground

clearance problems on a shallow sump by being mounted horizontally.

replacement for the stock filter. They're both available in short and long cartridges. However, the size of these cartridges is somewhat misleading because even the short cartridge will flow far more oil than an expensive paper filter.

One of the advantages of the System One filter over the Oberg is that its ribbed screen element has far more square inches of flow area and so pressure drop across the filter is minimal.

As far as the filtration capability of metal screen filters is concerned, it would seem that the metal screen filters are not able to screen particles as fine as paper screens. However, the particles that may pass through the metal filter may be too small to be of any consequence to the sort of bearing clearances involved in a typical small-block Chevy.

A typical stainless mesh screen filters down to about 45 microns. An average OE style filter is good to about 25 microns but bypasses so much potentially dirty oil this becomes almost academic.

One of the main advantages with the stainless screen filters, such as the System One unit, is that the flow

Here's a cutaway of the System One filter. The pleated screen gives an extremely large filtering area, far more than the flat-screen alternatives. Flow with this type of filter is typically between eight and ten times that of an average OE-type replacement filter.

capability is far higher than the pump's output capability, therefore it is practical to filter the oil 100 percent. For the System One, flow is about 8-9 times that of a stock OE filter. This means it can be used with a block adapter with no bypass. Since the filters can be stripped and cleaned, the other advantage is that it's there for the life of the vehicle.

A disadvantage of any metallic screen filter is that it cannot filter out water or acids from combustion as paper does. This means that oil changes must be more frequent.

Virtually all of my dyno test engines use System One filters. The reason is that they allow near-instant oil analysis after break-in by inspecting what has been stopped by the screen. Since these filters come apart, it is easy to see the debris collected on the filter element. On the other hand, opening up a paper filter is a pain. Also, attempting to inspect the fuzzy surface of a paper filter is difficult because it camouflages much of the debris. The stainless element does not do this. Since test engines are run for a relatively short time in terms of mileage, and are always on a diet of clean, fresh oil, the micro-particle filtration capability of a stainless screen becomes relatively academic as it would do in most cases on a race car. For street vehicles, where oil changes are at 3,000 miles rather than at about every thirty to fifty miles, I use K&N paper element filters.

Top-End Oil Restriction

A significant proportion of the oil that finds its way into the crankshaft's domain originates from the valve-train. Motors with the stock-type ball-mounted rockers need a good supply to the top end to prevent the ball and rocker assembly from galling and overheating.

If roller rockers are used, the need for a large quantity of top-end oil is eliminated as roller rockers require only a minimal amount of oil. Some oil will be necessary to cool the valve springs, especially if they are doubles incorporating a flat-wound damper. The damper spring performs its function by generating friction between the two springs, and friction generates heat. Without sufficient oil to remove excessive heat away from the valve springs, the springs will quickly

lose their temper, and the load they exert will drop correspondingly.

Oil for the lubrication of the upper end of the motor passes through the two lifter galleries, through the metering valve and the lifter, through the hollow stem of the pushrod, and out into the rocker. It drains back mostly via the drain-back holes at the ends of the cylinder heads, and from there it runs into the lifter valley. At this point the oil can pass back into the pan via several different routes. It can run out through the two holes behind the cam drive gear, or it can drain back through the holes in the lifter valley over the cam lobes, and through the drain-back holes at the back of the block.

If roller rockers are used, the oil flow can be restricted at any one of a number of points: at the back of the galleries at the point where it enters from the rear cam bearing, at the lifters or at the pushrods. Probably the best place to restrict the oil is at the lifters, but it's not always the most convenient, so the most popular method is to install a restrictor at the back of the oil galleries. This modification only applies to solid-lifter-type applications, as hydraulic lifters need the additional oil for their operation. Restricting the oil at the back of the lifter galleries means there's a lot less oil returning to the crankcase via any of the aforementioned routes.

If a roller cam is used, it is entirely practical, and indeed desirable, to plug the oil return holes at the center of the lifter valley. These holes usually serve to splash oil onto the camshaft to help with lobe lubrication, but with a roller follower this is totally unnecessary.

If a flat-tappet camshaft is used it's still possible to plug or put stand-off tubes in these holes, but only if relatively short mileages are intended. If you intend using a flat-tappet cam on the street, do not plug them as it will lead to increased cam wear.

If the situation allows the plugging of the holes at the center of the lifter valley, then we have to assume that all the oil will return to the crankcase via either the two holes behind the timing chain, or the drain-back holes at the back of the block. Generally, it's best to encourage the oil to drain back to the pan via the back of the block, rather than from the front. The principal reason for this is that oil passing down the front of the

With a fully roller engine, there's a need to restrict the oil flow to the top end of the motor. These oil galley restrictors from Moroso will get the job done.

If a roller cam motor is used, it's not necessary to drain any of the oil back through the galley holes. Blocking the holes off or installing stand pipes, as seen here, applies whether the motor is a dry- or wet-sump unit.

block must bypass the timing chain. From there, it's more likely to get entrained in the crankshaft than oil returning down the back of the block.

For a drag-race engine, keeping the oil drain-back in the front rather than in the back is no problem since acceleration forces drive it to the back of the block anyway. For a road-race engine, the acceleration forces will cause the oil to go down the back of the block, and under braking conditions the oil will drain down the front. Engine horsepower isn't needed during braking, so it is not super-critical from that respect.

In any event, it pays to make sure that the block drains the oil down the face of the block, underneath the timing chain, and on to the pan as rapidly as possible because coming back on the throttle may cause the oil that went through the front holes to be collected by the crank.

If roller rockers are being used, it pays to put a debris screen in each of the return holes in the block. Sometimes a rocker will break up and when it does, needle bearings will pass into the lifter galley. If these get to the oil

pump, the results are disastrous. Using epoxy resin, a screen can be installed in each of the four oil return holes.

If the ball pivot rockers are retained, then there is no need to install debris screens in the block. Indeed, with the amount of oil at the top of the engine, using a ball-type rocker, the screens may impede this oil flow back to the bottom end, unless they are totally open.

Rear Main Caps

One of the most expensive steel caps to buy, especially if you are sticking with a wet-sump setup, is the rear main-bearing cap, which of course is also the mounting point for the oil pump. Even if steel caps are used on the four other main-bearing stations, on high-output engines there can sometimes be a problem with breakage of the rear main-bearing cap. This

is more prevalent on a 400ci motor than a 350, as the 400 cap has a larger bearing diameter to accommodate. This leaves the material thinner between the bearing OD and the oil pump mounting face, and often results in cracked caps.

The cost of buying a steel rear main-bearing cap can be offset by reinforcing the rear main bearing with a steel plate. One provision must be allowed, however; all the oil pump mounting must be reproduced on the plate. To achieve this, mill down the rear main cap to an equivalent thickness of the rear main reinforcing plate. If you don't have the facility to make one of your own steel plates, then a suitable plate can be obtained from Ray Baker's Pro/Cam company. This makes for a stouter rear main-cap assembly, and has provisions for installing the oil pump in the stock position.

Dry-Sump Oiling Systems

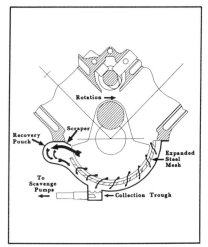

16

The layout for a dry sump, as opposed to wet setup, in terms of removing the oil from the rotating parts, differs in the fact that the oil is collected in an offset trough. From here it's returned to the reservoir tank by scavenge pumps.

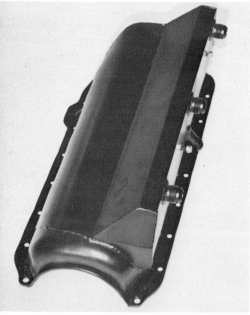

These two views show the design of a Pro Cam dry-sump pan. Note how the depth directly under the pan is minimized, and that the oil pick-up trough is off to one side. The three outlet pick-ups connect to three scavenge pumps.

Although wet-sump systems can be made to work under most circumstance, there comes a time when the effort and ingenuity put into their construction is simply not worth it.

A modern race car is capable of pulling high-cornering G's. High-cornering potential means that engine height becomes critical—the lower the engine is mounted in the chassis, the better. But this almost certainly means going to a dry-sump system. Ground clearance isn't the only consideration, however; a properly designed dry-sump system can show benefits in terms of power over a wet-sump system. The key phrase here is a properly designed dry-sump system.

Wet Versus Dry Sump

In a normal wet-sump system, the oil for the engine is kept in a reservoir below the crankshaft. Hence, it's necessary to strike a compromise between the depth of the oil and the installational height of the engine. Ideally, the sump for an engine needs to be extremely deep so that in spite of high-cornering G's, the oil never uncovers the pickup.

Unfortunately, a deep sump means either a high ground clearance or an engine set high in the chassis—neither of which is acceptable to many forms of racing where high-cornering G's are involved. Lowering the engine in the chassis to an acceptable level may result in the oil pan being too shallow for optimum effect. Even with all the trick baffling in the world, the lower limit seems to be about 3in below the lowermost rotating part of the crankshaft assembly. Having the option to lower the engine say, another 3in in the chassis, can have a significant effect on the cornering capability of most vehicles.

The only other option is to store engine oil in a remote reservoir rather than beneath the crank. Then you'll need to install a pressure pump to supply the lubrication system's needs, and scavenge pumps to return the oil back from the bearings to the reservoir. All dry-sump systems operate this way. Although the pan itself is not dry, it doesn't contain the reservoir oil, and if the system is good, it will contain no more than about a quart of oil.

Minimizing Power Losses

The use of a dry-sump system does not guarantee an increase in horsepower over a wet-sump system. A dry-sump system designed without adequate thought can lose as much, if not more, horsepower than a stock wet-sump.

By far the greatest source of horsepower loss can be bad pan design. Unless it's a really bad design, a wet-sump always has some space for the oil to go when it finally is centrifuged from the crank. However, with a closely fitting dry-sump pan there is a distinct possibility for a much larger quantity of oil to become entrained in the crankshaft and rotating parts. It is also possible to lose as much as 4qt. of oil in the rotating assembly.

The key to making a dry-sump system work is effectively removing oil from the crankcase. If it's a road or circle-track race pan, then obviously, a shallow pan is going to be required to allow the engine to be lowered as much as possible.

Important features of a pan are the use and position of scrapers to remove oil from the crank assembly, and the way the oil is directed from the scrapers into the recovery pouch. The recovery pouch takes the oil, turns it around and directs it toward the pickup trough. Scraping and directing the oil away from the crank also stops it from bouncing back into the rotating assembly. Because of the close proximity of the crank to the bottom of the pan, a solid screen is by far the most preferable.

Many oil pans have a centrally located pickup trough, but I feel it is better situated on the upward-moving side of the crank. It's also important to remember that oil has to drain to either the front or the back of the trough, so the floor of the trough needs to be high in the center with a scavenge connection at each of the lower ends.

If the sump is to be used for drag racing, an alternative approach exists. Here, ground clearance isn't a problem, so a deep pan can be used. A scraper to take the oil off the rotating assembly is, of course, mandatory, but because space is usually greater, the recovery pouch and drain-back trough can be of much larger proportions, especially depthwise. Also, the screen separating the crank from the scraped oil can be more effective when made of expanded mesh. This type of construction tends to reduce oil bouncing back into the crank, but works only in conjunction with a large space between the mesh and the bottom of the pan.

So far we've only discussed the installation of one pan scraper. It doesn't hurt to have additional scrapers, though. Because of the way the crank and rod assembly changes its form as the crank rotates, it's a good idea to try and scrape the oil off the crank at more than one position. With careful design, oil can be scraped from the crank over an angle of almost 90deg. Three and possibly four scraper blades can be accommodated.

It's hard to be specific in terms of potential horsepower gain from using a dry-sump system. Obviously the higher rpm the engine is capable of, the more potential there is for retrieving horsepower lost to windage and oil shearing. Most of my test engines turn up to about 7500rpm, with a few occasionally going to about 8000. I keep to relatively moderate rpm because most engines are required to survive for extended test periods. In practice, many drag-racing engines run far higher rpm than this. Even so, I've seen 10-25hp increases, depending upon the effectiveness of the wet-sump replaced.

Basic Oil Routing

Starting at the reservoir tank, oil is drawn off from the bottom of the tank by the pressure pump. After going through the pressure stage and being regulated to the desired pressure, oil is fed to the oil cooler via a filtration system. Once the oil has passed through the cooler, it's fed into the block at an appropriate place.

Let's assume the oil has gone through the engine and the bearings, has done its job and is now being collected in the trough of our highly developed dry-sump pan. The scavenge side of the system should have three pumps; it is possible to manage with two, but three are preferable. One complete stage will be used at each end of the trough in the pan. The third stage will be used to pull oil and air mist from the top of the engine.

Early ideas as to where the oil should be collected in the top end involved adapting the valve cover to take a large fitting. This would allow oil to be drawn from the back corner of the valve cover on the side of the engine which is predominantly on the outside of the turns. This is acceptable to a point, but it does leave oil from the other cylinder head to pass down into the lifter valley, and on into the bottom end of the engine, where it may be caught up in the crank.

A better idea is to drill into the outside cylinder head on the back face, into the area where the cylinder head drain-back hole is. Oil draining back into the drain-back hole is scavenged, but of course this still leaves the question of oil coming down the other side of the motor.

A third alternative is to block off all oil returns in the lifter valley which connects to the bottom end, and then put a pickup pipe at the back end of the engine. In this case, all the oil from both cylinder heads can be scavenged, and although more oil *may* collect in the lifter valley, the amount is almost academic because surplus oil here has little effect toward reducing horsepower.

Once oil has been scavenged from the engine it needs to pass through a high-flow filter. Remember, there is a limited amount of suction a pump can apply, and an overly restrictive oil filter will cause a major problem. Some engine builders use inline filters, and if the filter is sturdy enough, with sufficient flow this will work

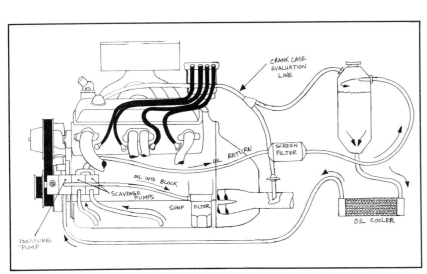

Here we're showing a schematic of the dry-sump layout in its more simple form. In this layout the oil input is in the same position as stock—the filter housing. There are other alternatives, as the text suggests.

A convenient point to block off the original and now redundant oil passage from the wet-sump pump is here, on the main cap, but there are many other different alternatives, as discussed in the text.

adequately—except that usually filters of this size, with adequate flow, only have enough filtration capability to minimally protect the scavenge pumps.

Any debris dropping into the engine is going to be immediately picked up by the scavenge side, so unless the filters can pull out the debris in the oil, the scavenge side of the pumps will suffer. If the particles are big enough, such as broken rollers, they will be destroyed. For most purposes, it would appear that a large System One filter in each line can handle this kind of job, so long as the fittings being used have large and smooth radiused turns.

After passing through filters and pumps, the hot oil-air mix is directed tangentially into the reservoir tank. This causes it to swirl around the side of the tank, and the centrifugal force developed de-aerates the oil. As oil drops down the tank, it should be passed through a screen to kill any orbiting motion and further de-aerate the oil. The hot and as yet uncooled oil is now ready to be circulated through the engine again.

Block Routing

When converting to a dry-sump system, it's necessary to reroute the oil's passage through the block. Remember first that the oil passages from the original wet-sump oil pump are no longer needed. The easiest way to fix this is to tap the oil supply hole at the block's rear main-cap interface, and plug it with a suitable Allen socket screw.

Chances are if you're going to a dry-sump system, you will also more than likely be replacing the rear main cap with a steel one. And since it's going to be dry-sump, there's obviously no need to provide for the wet-sump oil pump. Driving-in a suitable tapered brass plug flush to the block's surface will get the job done because once the main cap is on, the plug will not be going anywhere.

You now must decide where the oil from the pressure side of your dry-sump pump can be fed into the engine. The simplest place is to put it in with the stock filter setup, and the easiest way to do this is to obtain one of the many adapters available on the market, and then plumb in the oil. Though such a method is satisfactory, it is not necessarily the best. Every time oil takes a sharp corner, it's going to lose pressure. Even when the oil is hot it is possible to lose between 2 and 5psi at each turn, depending upon the acuteness of the turn.

In the block preparation chapter, we went to a lot of trouble to smooth out various sharp turns in order to maintain as much oil pressure as possible, without necessarily going to a high-output high-pressure pump. Putting the oil in by using an adapter at the filter mounting only marginally relieves the oil of having to make one or two additional turns.

A more popular method among engine builders for inputting oil involves eliminating some angles. Just above the filter housing is a plug usually requiring a square section key. By taking this plug out, and drilling the hole to accept a number 12 fitting (¾in), oil can be put straight into the galley, thus eliminating some acute turns in the adapter, and one right angle turn in the block.

A third possibility is to input the oil at the front of the block. Locate the eighth pipe plug tapped hole at the front of the block, just above the main oil galley. This is normally blocked, but may be used for an oil pressure fitting in some stock applications. By enlarging this hole, and threading it to take a number 12 fitting, oil can be passed directly into the main galley in the block with only one right-angle turn.

A fourth possibility is to put the oil into the block directly into the main galley at the flywheel end. By choosing the correct fittings, this can be the most free-flowing route of any.

Of the four systems mentioned so far, I recommend the latter. If you so choose, you can make a further refinement. Instead of plumbing all the oil into the engine at the back of the block, the output from the pump is split and a ½in line is routed to the back of the block, and another to the front via the oil pressure drilling modified to take a number 10 fitting.

Though minimal, the advantage of going this route is that the front and

It's practical to put oil into the block from the pressure side of the dry-sump pump here at the filter mount, but there are many more alternatives that provide less flow restriction.

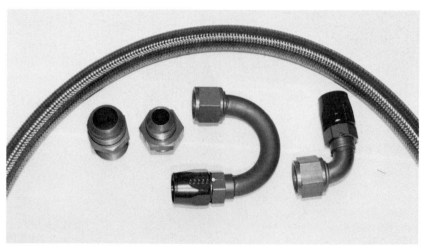

Abrupt turns in the oil line considerably cut flow. Larger radius tube fittings, like these from Goodrich, are considerably better than the cast right-angle joints often used.

rear bearings get oil of a more constant temperature. If fed in at one place the oil, which has had heat extracted by the cooler, will be at a lower temperature at the rear bearings than at the front. Under marginal conditions the oil at the front bearings may be too hot. If the rod bearings for the front cylinders seem to suffer more than any of the others, rerouting the oil in this way may cure the problem. For the most part, routing oil directly into the back of the galley produces minimal bearing problems under almost any race conditions.

Supply Lines

Anytime the dry-sump system is installed, it's a fair bet that the oil will have to travel farther to complete its circuit. Anytime an oil passage gets longer, it takes more pressure to drive the oil through that passage. Anytime passage becomes larger in diameter, it takes less pressure to push it through. From these statements, then, it follows that large lines should be used to convey oil around a dry-sump system as freely as possible.

This entails using ¾in bore lines —that's –12 braided hose. Design the line routing so that the run is as free of tight curves as possible. If you have to put in a union that has a right-angle turn in it, avoid using a tight right-angle bend; use unions that have a large radius tube between each fitting.

On the scavenge side you need to ensure that the line will not close up during suction. Some brands of braided hose are not as good as they could be in this respect. I've used genuine Aeroquip hose with success, but at the time I found them to be one of the most expensive brands. Earl's Performance Products was another brand I had overall success with, but since about 1987 I've been using lines from Goodrich because these have met all the necessary standards, and are considerably less expensive than most other brands, reliable or not.

Oil Pumps

A typical dry-sump pump setup consists of three or four individual pumps ganged-up together. One pump, or stage as it is often termed, is the pressure pump, and for a small-block Chevy usually delivers around 8gpm (gallons per minute) at 5000 pump rpm.

In the pressure stage there is also a pressure regulating valve, which allows the oil pressure to be adjusted. A good pressure to aim for in most instances is 55-60lb. Some top engine builders recommend 10psi per 1000rpm of engine speed. Though this is basically a good rule of thumb, it

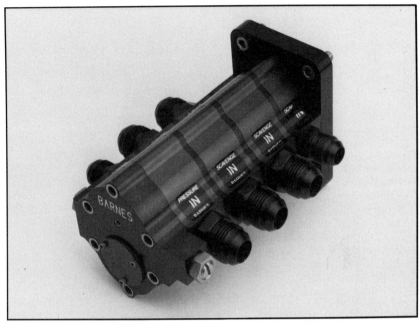

Here is a typical example of a dry-sump pump. The first stage next to the pressure relief valve adjusting screw is the pressure stage. The other three stages are scavenge.

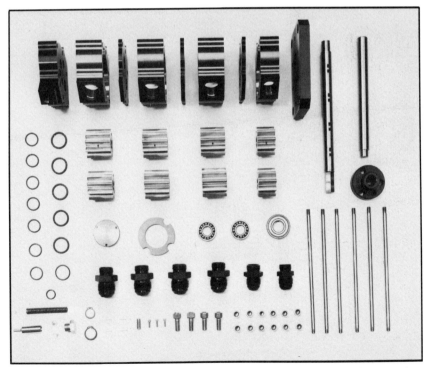

Stripped down, one of these pumps looks like this. As you can see this particular Barnes pump is of the gear type, though Barnes and other manufacturers, such as Peterson and Moroso, produce gerotor pumps.

145

does oversimplify the situation. It presumes that leakage from the bearings goes up proportionally to rpm, but in fact, this proves not to be the case.

The only time this rule might apply to any degree, is when there's a problem of oil overheating in the bearings. But there again, this should be cured by keeping the oil temperature a little lower, or formulating an oil that will withstand the temperature. Curing it by adding oil pressure means that there is a greater power loss driving the pump. Just remember, the higher the oil pressure the greater the quantity of oil circulated, and the greater the power drain on the pump. Also, excess oil pressure almost always leads to an oil control problem in the pan, thus aggravating windage losses.

Basically, the scavenge stages of a dry-sump pump are replicas of the pressure stage, except they have no pressure relief valve built in. Their purpose is to pick up oil from the pan and return it to the reserve tank. Although a small-block Chevy engine can get by with two scavenge stages, three are preferred, as mentioned. Two need to be connected to the pan, and one to the top end of the engine. An ideal situation with scavenge pumps is to have enough scavenge capacity to be able to reduce the pressure in the crankcase below atmospheric. This brings us to the subject of crankcase ventilation on a dry-sump motor.

Crankcase Ventilation

There are two schools of thought on the subject of crankcase ventilation. The normally accepted method, at least up until the early 1980s, was to vent the crankcase at as many points as possible. This allowed the scavenge pumps to freely draw the oil and air mixture through the engine.

The second theory is that the crankcasing needs to be sealed. In this situation when the scavenge pumps draw on the crankcase, they will pull oil and air out, thus reducing air pressure to below atmospheric. As the air pressure is reduced, the oil's ability to de-aerate itself is improved. This will help cut windage losses and reduce the possibility of oil finding its way past rings or valve guides and contaminating the intake charge.

To be able to successfully seal the crankcase, rings must function well to keep blowby to a minimum. If excessive, the blowby will outpace the scavenge pump's capacity, leaving you in the undesirable position of having a pressurized crankcase.

With many wet-sump applications an evacupan system is used, where the exhaust, going by the end of a tube, pulls a vacuum on a line connected to the crankcase. In this line there is typically an antibackfire valve, which closes if the line pressure goes above the crankcase pressure. Installing a pair of these will handle most of the crankcase evacuation, while the scavenge pumps are simply icing on the cake. Even if blowby due to ring wear reaches sizable proportions, the evacupan system can usually cope.

When crankcase pressures are reduced by more than about 30in of water, the oil seals may begin to leak air into the system. If this happens, then either a double back-to-back seal system is needed on the crankshaft, or the seals need to be reversed. A heavy-duty oil seal, such as the one available from Engine Tech, will usually take care of the problem.

To keep crankcase pressures low, Cosworth offers a rather unique dry-sump pump unit, in that it has a Rootes-type supercharger scavenge assembly. This is capable of removing some 8cfm of air and oil mist. It also allows a substantial vacuum to be pulled on the crankcase. Typical blowby on a race engine should not exceed about 3cfm, so the pump's capability should far outpace it. Thus, the potential exists to drop the crankcase pressure.

After having gone through this mini supercharger, the air and oil mist is fed into a centrifugal oil-air

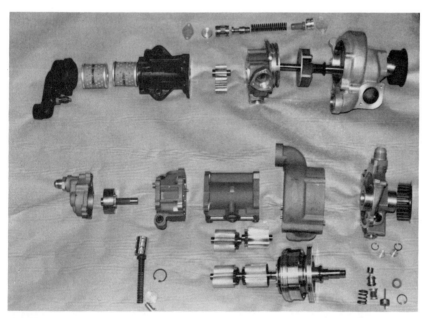

Here is an exploded view of the Cosworth combined water-pump and dry-sump system. Even though the price is somewhat steep, as you can see from this exploded view a lot of parts go towards making this pump system as effective as it is.

This centrifugal oil separator in this Cosworth system ensures that the scavenge pumps handle mostly oil, rather than oil and air.

separator before being returned to the reservoir tank. This Cosworth pump, which can also incorporate a remote water pump, is a derivative of the one used on the Indy and Formula One engines. It is also becoming popular with the leading teams in road racing. It represents a significant improvement in dry-sump state of art. Since de-aeration at the reservoir tank is never 100 percent effective, the inclusion of the centrifugal air-oil separation has resulted in improved bearing life on highly stressed endurance engines.

Excluding the Cosworth pump, the pumps currently available fall into two categories: the gear type, and the gerotor type. Gear-type pumps predominate in US-built dry-sump systems. For a given pumping capacity they can be made a little smaller than gerotor-type pumps.

My experience indicates that, for a given capacity, gerotor pumps are more efficient because they absorb less horsepower to drive them. I hesitate to state exactly how much more efficient they are, though, since so many variables are involved. A realistic estimate is 30 percent.

On the debit side, I have heard that some gerotor installations have given problems on the scavenge side. I have not experienced problems, but some of Chevy's IROC race cars had problems back in the mid 1980s with premature bearing failures. After much searching it was decided that the probable cause was that the gerotor pump was causing excessive aeration in the return. Since that time changes have been made to many of

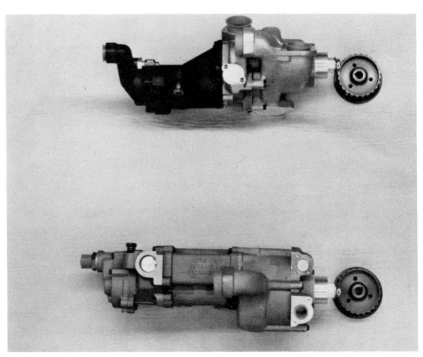

This is the complete installation for the Cosworth setup. The upper section handles water and the pressure side of the lubrication system, while the lower part handles the scavenge side of the lubrication system.

This miniature Roots-type supercharger in the Cosworth system ensures that crankcase pressures are drawn down well below atmospheric.

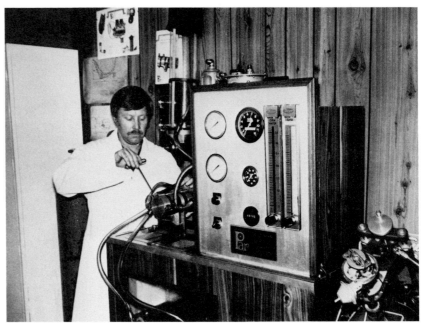

Here is the pump dynamometer used at Peterson Oil Pumps. Basically it measures the current draw on an electric motor for a given amount of oil supply.

Tests of gerotor pumps versus gear pumps show that gerotor pumps typically take a substantially less amount to drive them.

the gerotor systems. Their common use in Europe shows they can work so maybe it's just been a case of debugging for use on the small-block.

Gear pumps are the most common type used for small-block Chevys, and the argument for their use appears to be that they can ingest engine debris easier without suffering terminal damage than a gerotor type.

In Europe, however, there are more race engines with gerotor-type pumps; they probably outnumber the gear-type pumps by more than 100 to 1. This ratio is just about reversed in the United States on small-block Chevys. Granted, the gerotor type may not be able to ingest particles as easily, but there is no justification for not filtering the oil prior to the pump —regardless of the type of pump involved.

Pump Mounting

Another point to consider when looking for a pump setup is the mounting. A good, stout mounting system is important. Most pump manufacturers take care of this by providing an adequately stiff single-thickness mounting plate or dual mounting plates, depending on the position intended for the pump. If you are mounting the pump in a special location, and need to make up custom brackets, be sure the pump is mounted securely.

Oil Filters

The most important side of a dry-sump system is the scavenge side. The most important part of the filter is the pressure side. Although this is obvious, often expensive engines are put together that are inadequately protected on either side.

When filtering oil for a dry-sump motor, you need to be aware of two things: First, the filter capacity for dealing with the scavenge side must have plenty of flow capability. If it doesn't, it can seriously impede the performance of the dry-sump system. Plenty of inline filters are available, but many seem to be marginal in their performance.

If you intend using an inline filter I recommend the System One units, as they have 65sq-in of filter area for minimal restriction. Though more cumbersome, I also recommend using a large System One filter in each scavenge line for optimal results. For ease of inspection, remember to connect them up so that the debris is left on the outside of the stainless screen mesh element. Because the System One element is pleated, it has a much larger filtering area than a flat screen

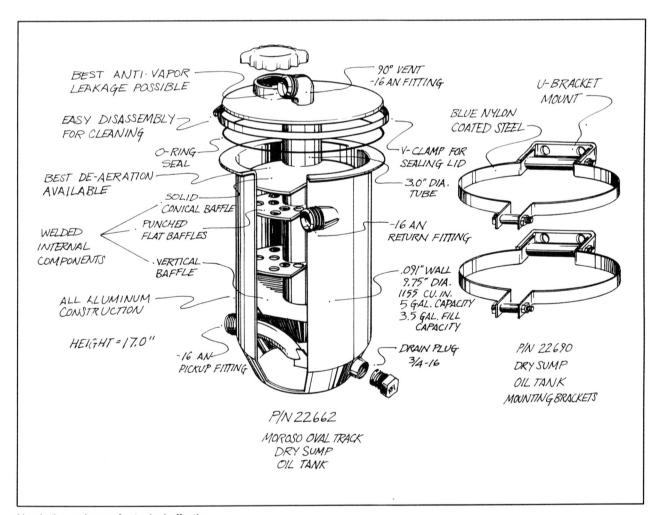

Here's the makeup of a typical effective dry-sump tank; this particular item is available from Moroso.

filter, and so has greater flow capability and takes longer to clog.

The most important time to check the filters on the scavenge side is immediately after the engine's break-in. It is amazing how much shop-rag lint and paper towels can accumulate during assembly. Such debris goes almost unnoticed on a paper filter, but shows up on a metal screen filter. By and large, the presence of paper or even fine lint from shop rags, which is of minute proportions, does not cause any harm to the engine. On the other hand, grit does, so heavy use of paper towels during assembly can never be construed as bad practice. Lint from paper towels can block up wire screen filters, though. But once the initial oil change for the engine has been done, and break-in is completed, the only debris you're likely to see in the scavenge side will be particles of metal from normal engine wear, or parts that are breaking up that shouldn't be.

Oil Coolers

Although some engine builders allow oil temperatures to rise as high as 270deg. Fahrenheit in a wet-sump, I am not particularly enamored with such high oil temperatures. My own dyno testing has shown that oil temperatures significantly over 220deg. do not contribute to further increases in power output. On the other hand, using up the margin between a safe oil temperature and an excessive oil temperature seems to be a fruitless exercise. If keeping the oil temperature down to 220deg. means that there's an extra margin for unforeseen events, then that's obviously the thing to do.

I prefer to place the oil cooler on the pressure side of the system, immediately prior to the oil's passage into the engine. Several advantages are gained by placing the cooler here. First, the oil pump is always pumping hot oil. This means that it is less viscose, and easier to pump.

On numerous occasions, I've seen oil coolers installed in the scavenge side of the system. This has several disadvantages. One is that it requires a greater amount of oil cooler area because the scavenge pumps are putting not only oil, but also air through the cooler. There is no future in cooling air. The easiest way to get rid of the hot air is to simply dissipate it, mostly through the evacupan system, or through the vent in the reservoir tank.

Another disadvantage is that the oil viscosity will increase due to the cooling action prior to it going in the reservoir tank. The reservoir tank de-aerates the oil mostly by centrifuging it. Trying to centrifuge thick, cool oil is harder than thin, hot oil. Also, bubbles will take longer to dissipate out of thick oil than they will out of thin oil, so putting the cooler on the scavenge side just makes life more difficult all around.

Reservoir Tanks

There are plenty of good reservoir tanks available on the market, so there is no real need to make your own. Look for the tallest tank possible; it should be capable of holding at least 8qt. of oil when approximately two-thirds full.

The tank needs to be adequately vented to the engine, produce a good centrifuging action as it puts the oil into the tank, and have the pickup at the bottom of a cone-shaped lower end.

If the car is to be used predominantly for circle-track racing, where high lateral G's one way are seen, then it's not a bad idea to incline the tank about 18-20deg. This will reduce about half of the cornering G's in the turn, but give the effect of increasing G's the other way down the straight, so the tank will average about the same G's both ways, and the surface of the oil in the tank should remain much nearer the horizontal. Under these conditions, the centrifuging and the oil-air separation action as the oil drops through the grid will be the most effective.

Of course, if the engine is to be used in a road-racing setup, then a substantial amount of cornering G's either way can be seen. But it's still predominantly more one way than the other due to the track's overall rotational direction.

Proofing the System

Since there are many variables in putting together a dry-sump system, it's necessary to proof the system, especially if you've never constructed one before. This is best done on a dynamometer, although these days it could be done with a video camera at the racetrack.

With the engine on the dyno and idling, take a look at the amount of oil in the reservoir tank. You'll need a sight tube on the tank so that it's possible to see exactly where the oil level is. Now run the engine up to about its maximum rpm, and check to see how much the oil level drops. The amount it decreases will be the amount of oil retained in the pan. A quart is typical of an acceptable system, although a good system may retain only half this amount. Significantly more than a quart remaining in the pan indicates that the crankcase is not being scavenged properly, and a potential power loss is occurring.

If you don't have access to a dyno, using a video camera and filming what happens to the fluid level in the tank will achieve the same purpose. Bear in mind, though, with most installations the reservoir tank is usually in the trunk if it's a road-race type of machine, or if there is plenty of room up front, it could be under the hood. In either place, it's not visible to the driver, hence the need for a video camera. If too much oil stays in the pan, it's more than likely that the scrapers are not doing their job. But you also need to consider that the oil pressure may be too high, or bearing clearances too wide.

System Priming

When you are about to fire up your newly built dry-sump small-block Chevy for the first time, prime the oiling system first. This is easily done by removing the pump driving belt, and using an electric drill to run the pump until oil pressure is achieved.

Cooling Systems and Water Pumps

17

It's easy to assume that since a stock cooling system doesn't give undue problems, nothing needs to be done to improve it. If all you're interested in is a stock motor, fine. But if you're making performance modifications that will result in more horsepower you will need a better cooling system as well.

Cooling Requirements

Three factors must be considered for an ideal cooling system. First of all, the system must allow sufficient temperature to be developed so that heat conduction from the burning charge to the surfaces of the cylinder walls and combustion chamber isn't too great. If too much heat is pulled out of the charge, the engine will lose thermal efficiency as the pressure generated during expansion will be less, and consequently so will power.

The second factor is at the other end of the thermal scale: temperatures must not be too high, otherwise there will be an early onset of detonation. The compression ratio will also be limited, and the additional excess heat will cause a charge density reduction, and thus lower the volumetric efficiency of the engine. All of this will bring about a reduction in potential power.

A final factor to consider is that the engine must have a uniform overall temperature. Obviously if the engine suffers from hot spots, these may limit the situation before the overall temperature does. If one cylinder starts to detonate due to a hot spot in the chamber because of inadequate cooling, it becomes the limiting factor even though the water temperature may not have gone above normal anywhere else. Achieving a uniform temperature in a small-block Chevy is difficult. The problem also differs between nonsiamese and siamese bore blocks because the water circulation patterns are different.

Coolant Flow

Before moving along to any modifications, we need to study what the basic coolant flow through the block and cylinder heads should be.

Cool water is drawn from the radiator by the water pump. From here it is pushed into each bank of cylinders from the front of the block. Then it should pass down through the block and up through larger holes at the back of the block into the cylinder heads where it moves forward through the heads and out through the manifold thermostat passage. From the thermostat, which regulates the flow according to temperature, it passes back into the radiator. The water is of course coldest at the front of the block; as it goes through the block it picks up heat from each of the cylinders, so at the back of the block the water is hotter.

This hotter water goes up into the cylinder heads where it moves from the back to the front. If this route is used repeatedly, the numbers one and two cylinder bores are likely to be the coolest while numbers one and two combustion chambers are likely to be the hottest.

To compensate for this there are communicating holes between the block and the cylinder head. As water moves through the block from front to back, some of it is redirected up through the block into the heads. This cooler water mixes with the hotter water moving from the back of the heads and ideally evens out the temperatures. Though the system may prove adequate for a normal street machine, it certainly is far from satisfactory for a race engine or indeed even a high-performance street engine where maximum power is sought.

The problem is that the water does not circulate through the block and cylinder heads as simply as expected. In addition, the water itself is not necessarily a perfect coolant. If any particular part of the engine has a greater amount of heat to dispose of and the circulation is low in that area, the water can flash into steam locally and exaggerate the hot spot. This may mean that though the bulk water temperature is below what we perceive as a dangerous level, it may be well above what the engine can tolerate, bearing in mind the compression and fuel it is using.

Under such conditions, the power output of the engine will be limited because of a localized hot spot. Additionally, localized hot spots, especially with aluminum heads, can result in cylinder head failure. Fortunately, to a large extent, most aluminum-head manufacturers have built cylinder heads with deck thicknesses strong enough and with sufficient water capacity to attempt to compensate for any idiosyncrasies that may be in the block. Obviously when designing a new cylinder head, it won't be necessary to compensate for existing cylinder head cooling problems—one can simply design them out in the first place.

My experience has been that cooling problems occur less frequently with aluminum heads than with iron heads. Of course, the added conductivity of aluminum helps. Most enthusiasts building small-block Chevys will be using iron heads, however, so we will need to address these problems.

Flow Modifications

Since the cooling system doesn't work as intended, the engine has far from uniform cooling temperatures and combustion chamber-cylinder wall temperatures. In practice, the hottest and coldest cylinders can vary somewhat depending upon the type of castings used and variations between one casting and another of the same type. On a high-performance small-block intended for the street, it would appear that the number two cylinder runs coolest while cylinders three and five run hottest. This is partly due to the fact that the water pump does not discharge water evenly.

The design of a typical water pump dictates that more water is discharged into the left-hand side of the block as you're facing the front of it. By simply blocking off half of the left-hand outlet from the water pump, this will restrict the flow to the overcooled side and force more water to the other side of the block, thus bringing the temperature down from cylinders three and five. Incidentally, cylinder number five is usually the first to detonate if the mixture to all cylinders is uniform. Of course this detonation can be affected by manifold distribution also, and the temperature can often be a primary indication of bad fuel distribution.

Having partially blocked off the water pump outlet to the even-numbered cylinders, there is still the question of flow distribution in terms of what effect the bypass drilling has on the temperatures on the even-numbered cylinders. There's a small hole situated directly below the pump outlet hole. When the major water-pump outlet hole is partially blocked off, additional flow is diverted to the bypass. This source of coolant flow is not required and is best dealt with by plugging the communicating bypass hole in the block face. The easiest way to deal with this is to tap the hole out and install a plug before the block is decked.

Earlier we discussed in detail the preparation of siamese bore blocks and the plugging of various holes in the deck. To some extent this can also apply to a 350ci block. Some engine builders will plug the large holes and then redrill them with a small hole. Plugging the larger holes is probably more beneficial in terms of prestressing and stabilizing the block deck than any effect it may have on cooling—these holes are large in the block, but they are usually relatively small in either the gasket or the cylinder head. As a result the large hole is not the controlling factor in terms of water flow.

Water flow on the inside edges of the block, between the block and head, is not as critical as it is on the outside. Remember, the inside part of the cylinder head has only cold intake ports to deal with, whereas the outer parts have the hot exhaust ports to cool. We need to encourage water to flow down the hot side of the cylinder head a little more than it does down the cold side. This will even up combustion chamber temperatures, and can also contribute toward better spark plug boss cooling.

Keeping spark plugs cooled, or at least allowing them the ability to be cooled and then selecting the heat range to compensate, is a good idea to get plug temperatures where you want them rather than where the engine forces you to use them. Additional cooling of the number two boss may not be necessary because it can still be the coolest-running plug area out of the eight. However, it is wise to pay some attention to the flow of water around the cooling bosses on cylinders three, five, seven and eight.

Another area that can utilize better cooling is where the two exhaust valves are adjacent to each other in the center of each side of the block—that's between three and five, and four and six. I haven't yet found an effective way of increasing the cooling at this point by drilling additional holes in the block face. With siamese blocks, I recommend drilling a transfer passage parallel to the deck face, as described earlier. Other than that, it appears the only way to aid the cooling of these two areas is to improve the overall effectiveness of the cooling system.

Some engine builders drill additional holes in the block so that more water can come up from the block into the cylinder. I've tried this with limited success, especially on 350s. In some instances it has marginally helped, especially on the deck face between the hotter-running three and five cylinders. Since variations from block to block can often significantly alter the water-flow patterns, you can never be sure that the modifications are actually effective. Only when a large number of blocks are modified and the average results viewed can we be positive of the success rate. Drilling additional holes doesn't seem to harm an engine, however, so it may be worthwhile considering.

Controlling Flow

Probably the most effective and simple add-on piece to improve the cooling system is the Moroso Y-outlet top water takeoff adapter. This does away with the standard thermostat exit point, and instead of taking the water from the front two holes in the manifold face of the heads, water is drawn off from the cylinder heads by holes tapped in the front face of the heads. By drawing the water off from the lower side we can increase the flow down the hot exhaust side of the head more so than down the colder intake side. This has a tendency to cool the cylinder head more evenly. It also allows a greater flow of water.

This particular piece of equipment, sold by Moroso, was developed by Smokey Yunick. Yunick's explanation of how it works is relatively straightforward. During one of his many test sessions, he found that the temperature of either bank of cylinders would cycle. First the temperature would rise and then it would drop, then another rise and drop. So he searched for the mechanism causing this, and concluded that with the two water channels meeting at the thermostat, interference was taking place that caused a cyclic flow of water in each bank of cylinders.

As one cylinder got hotter, the pressure in that side of the block increased, thereby reducing the flow at the thermostat on the cooler side —the pressure on the hot side, of course, overcoming the cooler side. This would cause the water in the cool side to stay in the block longer and pick up more heat, whereas on the higher pressure hot side, the water would be encouraged to flow through slightly easier. As a result, the cooler water would flow into the hot side and cool it, while the cool side raised in temperature. Once the cool side had become hotter than the hot side, the flow would reverse, and this would go back and forth ad infinitum. By pulling the water off from the cylinder head, and bringing it together in a more streamlined form, this cyclic variation was reduced and, of course, the flow increased.

At this point let's discuss a few fallacies of cooling systems. First of all, there is no such thing as getting the water through the system too quickly. Some engine builders argue that if the water goes through the system too fast it's not in there long enough to reduce the heat. What they overlook is that if the water goes through twice as fast, it goes through twice as often, so it has twice as many chances to remove the water.

Faster flowing water also causes more turbulence and it is a proven fact that turbulence in a heat transfer system is required to improve the efficiency of the system. Highly turbulent flow inside the blocks and cylinder heads in radiators tends to scour the surface and remove the hotter boundary layers adjacent to their surfaces. Experiments may indicate that if there is no restriction in the system, the engine runs hotter, thus the argument that if the water goes through

151

too quickly, it doesn't reduce the heat. That is not the situation, however.

The water pump can develop high pressure in the block. If the outlets from the block are completely stalled, the water will develop its maximum pressure. On the other hand, if the water flows through the block effectively, less pressure will develop in the block.

With all the casting imperfections and nooks and crannies within the system, there are plenty of places that develop hot spots. When these hot spots form, water at the surface turns to steam. Once this happens, the hot spot becomes even hotter. The point at which it turns to steam depends upon the pressure of the system. If, as is usually the case, the system is pressurized by means of a 15psi pressure cap, the boiling point of the water will be raised to about 250deg. F.

The pressure in the block, generated by the water pump, adds significantly to this and it is possible to develop another 35-40psi on top of the pressure developed due to the heating of the water. This extra pressure increases the boiling point of the water locally, so if any water was about to form a steam pocket, it has to do so at a much higher temperature. If the water can flow through the block without developing pressure, then this can encourage the formation of steam pockets, which ultimately lead to the symptoms of the engine overheating.

To an extent, then, pressure in the block, as well as flow, can be relatively critical. For this reason, the Moroso Y-outlet has to have a certain amount of restriction built into it because it will flow a far greater amount than the stock thermostat and associated water-routing setup. It will not be necessary for the Y-outlet to have such high pump-generated coolant pressures as the stock system, however, because it allows a greater quantity of cooler water to flow into areas where the heat needs to be extracted.

Even if the stock cooling system was adequate for the job, this will allow a reduction in pump speed which will, in turn, require less power to drive the pump. With a regular water-glycol cooling system, the small-block Chevy needs to have a pump typically running at about 75 percent of engine speed.

Rerouting Flow

To improve the block cooling further, more radical modifications are necessary. The simplest of these is to pull water off from the pump and feed it directly to the back of the block so water now goes in both front and rear. This allows the front cylinders to run slightly hotter than they would normally and the rear cylinders to run cooler, which evens out their temperatures more. But it still leaves cylinder heads at something of a disadvantage because the water is generally flowing up into the heads and out the front.

The front combustion chambers can now run hotter than they would normally. If enough water is fed to the back of the block to even out cylinder temperatures we find that the combustion chambers near the front of the block may run too hot. To compensate for this, put the water into the block in a number of places. On siamese-bore blocks the freeze plugs can be removed, the holes suitably tapped and adapters fitted that will allow the water to go in on these three points on each side of the block. By pumping the water in here, generally the flow will be from the bottom of the block back up to the top.

To match this kind of water flow, pull the water out of the heads at various locations. Probably the best way to do this is to allow water to come from the two end passages, plus an additional center passage; some cylinder heads such as certain Brodix models already have center takeoffs tapped and ready to use.

Going this route means a water-flow system that tends to be up through the block to the cylinder heads—still not ideal for most purposes, as the coolest water should get to the hottest parts first. More even temperatures can be achieved when this already heated water is taken to the cooler parts. Simply put, most cooling systems appear to be running backward.

The problem with running the water through the block in the reverse direction is that a pump will tend to cavitate much easier in hot, rather than cool, water. In hot water the pump is almost to the point of causing it to flash to steam. Running the pump in water causes a pressure drop in certain areas of the blades. Where this pressure drop occurs, the water will flash easier to steam and therefore exhibit the typical characteristics of cavitation.

Ideally, what is needed is a cooling system that utilizes a coolant with a boiling point far higher than the engines' typical operating temperatures. This would allow a greater temperature margin between what is seen on the gauge and what occurs in various low-flow pockets within the water jacket. Fortunately, an alternative practical coolant exists.

Alternative Coolants

One of Rolls-Royce's most famous engines, the Merlin, was instrumental in the success of two of the Allies' best World War II fighter planes: the Spitfire and the P-51 Mustang. The fact that the Merlin was a powerful engine was only part of the story. Being a V-12 it also had a minimal frontal area, an important factor for a fighter plane where speed is of the essence—the more frontal area the engine has, the more power it needs to push its way through the air.

Being a liquid-cooled engine, the Merlin needed a radiator somewhere in the system to keep its operating temperature within safe limits. It's well known that the higher the temperatures being dealt with, the smaller the radiator needed to dissipate a certain quantity of heat. The problem is that a water-cooled engine cannot operate above the boiling point of the liquid, so Rolls-Royce elected to use straight glycol, which boiled at a much higher temperature. This allowed them to operate the cooling jacket of the Merlin at a higher temperature and thus get away with a much smaller radiator hanging out in the wind.

At first, running the engine at a higher temperature may look to be a disadvantage in terms of power. However, the situation is not quite as simple as it may seem, for there are a number of interrelated factors other than just the bulk temperature of the coolant. To see how some of these related factors affect the way the cooling system works and its subsequent effect on power, we need to understand a few basics about the way a typical water-antifreeze system works.

First, it's a misconception to think that the water simply moves through the jacket picking up heat. This kind of heat transfer occurs only during the initial phases of a warm-up period. As the engine starts to approach its operating temperature, nucleate cooling starts to take place. This level of cooling can be recognized by the small bubbles of high-temperature water vapor that develop at the surface being cooled. When they reach a certain size, they are carried away by the stream of water. Depending on the temperature of the bulk liquid, these bubbles of high-temperature water vapor may or may not condense and disappear.

The nucleate cooling phase can be likened to a pre-boiling phase. The water, in contact with the hot inner surfaces of the cooling jacket, is just hitting boiling point, and each bubble of vapor carried away from the surface from which it was generated is recovered with cooler water. Then the process repeats itself.

Since water absorbs a lot of heat as it goes through the boiling-vaporization phase, a great deal of heat is carried off. This type of cooling can be seen in the bottom of a saucepan on a heating element. It is the phase where small bubbles rise up from the hotter surface of the saucepan, but it occurs substantially before the main mass of water reaches the boiling point.

If the temperature of the coolant continues to rise, then the amount of steam generated at the hot surface can actually blanket the surface, shielding it from the fresh supply of lower temperature coolant. If a big enough surface flashes to steam, that surface will continue to rise in temperature as it is now protected from the coolant by the blanket of steam. This is especially true if the steam is formed in a pocket where it cannot easily be carried away by the flow of the coolant. There are numerous places in an engine's water jacket where this can occur.

When this cycle of cooling takes place, the hot spots within the engine become aggravated. Inevitably this will lead to an earlier onset of detonation which ultimately becomes the limiting factor toward extracting power from the engine. With optimum coolant effectiveness, the bulk operating temperature is as far away from the boiling temperature of the coolant as possible. When using water as a coolant, even under pressure, bulk temperature of the coolant must never go beyond 250deg., as this is the approximate boiling point of a pressurized cooling system.

If a coolant has a substantially higher boiling point than the normal operating temperature of the engine, two things occur: the coolant pump is far less likely to cavitate, and the water jacket is more likely to operate at a more even temperature. If the temperatures are evened out, then there's a possibility that the bulk coolant temperature can be raised without necessarily lowering the detonation limit of the engine. Remember, it's localized hot spots that inevitably lead to the initial onset of detonation, not the average temperature of the engine.

Water has one of the highest specific heats of any commonly used cooling liquid. As such, it is able to carry away a substantial amount of heat without necessarily having a high flow. Other coolants have to operate at a higher temperature to be able to carry away the same amount of heat for a given flow. Either this or such coolants would have to be pumped through the engine faster.

Propylene Glycol Coolant

Jack Evans of Meca Oiling Systems, Inc. has been researching effective cooling systems, utilizing a coolant that would show superior results to water or water-glycol mixes. He developed a coolant mix largely based on propylene glycol.

The advantages of propylene glycol are that it boils at approximately 370deg. F. at sea-level air pressure. Because the boiling point is substantially higher than the operating temperature, this means that cavitation at the pump is less likely to be experienced. In practice it would appear that in a stock-block, stock-manifold configuration, cavitation ultimately limits the flow of water through a small-block Chevy to a little less than 70gpm (gallons per minute).

By opting for propylene glycol, not only does more coolant pump through the engine, but also there's an increased possibility of successfully reversing the flow so that the coolest liquid reaches the cylinder heads first and then goes to the block, thus evening out some of the head-to-block temperature differentials.

Making a system work with propylene glycol involves more than simply reversing the flow and pouring in the liquid. Because of propylene glycol's high viscosity and its low specific heat, the pump drive ratio has to be increased to get flow through the block at rates consistent with the amount of heat that has to be carried away.

Most water pumps in high-performance applications are typically underdriven 10-30 percent. With the Meca propylene glycol system, water pumps need to be driven at about 90 percent overspeed. This means it's turning about twice as fast as it would with a water-based coolant system. The need for extra speed will, of course, absorb a little more horsepower. However, the

---

Detonation Test: Propylene Glycol Versus Water and Ethylene Glycol Mix

Coolant Out F	100% PG Detonation	50/50 H^{2}O & EG Detonation
190	No	No
200	No	Small
210	No	Medium
220	No	Large
230	No	Large
240	No	Large
250	No	Large
260	No	Large

Here are the results of a test performed by Meca on a late-model aluminum-headed Corvette engine. Using the regular mix of ethylene glycol and water, the engine developed trace detonation when the coolant outlet temperature reached 200deg. F. When this temperature rose

to 210, the engine was into moderate detonation and at 220 the detonation was severe. By changing the coolant to 100 percent propylene glycol, and reversing the flow through the engine, detonation was eliminated, even at 260deg. F.

benefits of the reverse-flow cooling system, such as the 1991 LT1 uses, have been shown in tests, and more than compensate for this.

To begin with, although the radiator needs to be deep to accommodate the amount of flow, its frontal area need not be as big as a water radiator. Because it's operating at a higher temperature, it can dissipate heat at a higher rate per square inch than a conventional water-cooled system. On a vehicle intended for high-speed use this can be important—it means that either frontal area or drag, or a combination of both, can be reduced to allow more horsepower to propel the car along rather than overcoming drag.

Although overspeeding the coolant pump with a propylene glycol system looks like a black mark against it, we have to consider the effectiveness of the overall concept. Propylene glycol allows a combination of higher temperatures and compression ratios to be used. In a test on a stock Corvette engine, the coolant temperature was raised from 190 to 260deg. F., while the propylene glycol hybrid cooling system completely suppressed detonation from the engine. The water-ethylene glycol system allowed the engine to detonate when the bulk temperature of the coolant was 200deg. F.

The next Meca test results worthy of mention are from a dyno test on a 337ci small-block with a compression ratio approaching 16.0:1. The average horsepower went up as the temperature was raised from 190 to 260deg. F. On the other hand, the average cubic feet per minute consumed by the en-gine dropped, while the average exhaust temperature climbed by almost 170 deg.

From these figures you can infer that there was an increase in horse-power—small, but nonetheless relevant. This power increase was seen at temperatures as much as 90deg. above those temperatures normally used to produce peak power with a water-glycol-cooled system.

From this you can also conclude, with a reasonable degree of certainty, that the propylene glycol is holding down the average temperatures of the cylinder heads. So, hot spotting is not as prevalent as it is for the water system. The fact that the bulk temperature is up is not the critical issue; it is the temperatures of the hot spots around the combustion chambers that are important. Whether a propylene glycol or conventional system is used, this points to the fact that evening out cylinder head temperatures is an important part of modifying an engine if it is ultimately to deliver the maximum power.

In this test, the engine used 17.7 less cfm to make a slightly higher power output. Why did the airflow decrease? Since the overall temperature of the engine increased, the air going into the engine is preheated more prior to arriving at the cylinder. Because it's preheated, it expands and this causes the volumetric efficiency of the engine to drop. However, the fact that the engine makes more horsepower on less air needs to be explained. Presumably, two factors apply here.

First, friction is reduced at slightly higher temperatures, so this could be one contributing factor. The other factor, probably the most significant, is that heat conduction from the combustion process through the cylinder walls is reduced because of the higher average temperature of the cylinder walls. This increases thermal efficiency and allows the charge to stay hotter longer, thus producing a higher average cylinder pressure. The net result—more horsepower.

Although the brake specific fuel consumption figures are not known, if additional horsepower is being produced on the same amount of fuel, or even less fuel if the airflow is reduced, then the fuel efficiency of the engine must have gone up. With the difference made, such a small change could not be expected to show up until the third decimal place.

A final factor may be the exhaust gas temperatures. Higher exhaust gas temperatures are a good indication that less heat is being conducted from the charge, and this, again, points toward the heat conduction theory as to why more horsepower is being developed. Apart from the increase in power, we also need to consider that this 16:1 motor was run at 260deg. F. jacket temperature without detonation. There is little chance of doing this with a conventional coolant system—unless the fuel octane is in the 110 plus range.

To see how the Meca hybrid cooling system functions, it helps to look at the surface temperatures of the metal. Meca's test results on an aluminum-headed Corvette engine serve as a useful demonstration. From the test results, at any given degree of coolant temperature, the hybrid system has much cooler metal temperatures. Some of the temperature changes have been substantial. For instance, the temperature between the middle cylinders—where both exhaust valves are adjacent—was hotter with a water-glycol coolant at a bulk temperature of 190deg. F. than with the propylene glycol system at a 270deg. bulk temperature. In fact, at this critical point we're talking about a temperature reduction of some 80-100deg., even though the coolant temperature was 80deg. higher.

Thus we can reasonably assume that the propylene glycol is much more effective at controlling the heat, but that's not all. Tests show the propylene glycol engine to be clear of

Power Versus Temperature With Propylene Glycol

Temp F	190	230	260	Change
Average HP	477.3	478.4	479.0	+1.7
Average BSFC	0.52	0.52	0.52	0
Average CFM	797.2	790.4	779.5	−17.7
Average EGT	1154	1297	1322	+168
Observed Knock	0	0	0	0

A test performed at various water outlet temperatures with 100 percent propylene glycol coolant and a reverse flow system produced these results. The most important aspect to consider here is that with a normal cooling system utilizing water and anti-freeze of some sort, maximum power temperatures tend to be in the region of 170-180deg., and rarely above 190. Here we see that the more even cooling given by the propylene glycol reverse cooling led to power continuing to climb right up to 260deg. F. Admittedly the changes are small. Note the drop in airflow, and hence volumetric efficiency, with the higher temperatures. From this data, you can only assume that friction and viscous losses, which traditionally drop with temperature rise, allowed the engine to make more horsepower.

detonation, even at 270deg. bulk temperature, whereas the water-glycol-cooled engine suffered detonation starting with a trace level at 220deg., and by the time it had reached some 250deg. coolant temperature, the detonation was severe.

By putting the engine farther from detonation it can utilize a spark curve that may be more optimal, whereas temperature and octane levels may limit the amount of spark advance that can be used in the conventional cooling system. A near-optimal spark advance is a practical possibility when the propylene glycol cooling system was used. By being able to optimize the spark, a considerable amount of extra torque and power is then available for the engine throughout most of the rev range.

Although it has proved effective, the reverse-flow cooling system from Meca does require a substantial amount of hardware to implement. Apart from special water pump, special head gaskets to redivert the flow are also necessary. Since these tests were done, Meca has moved forward

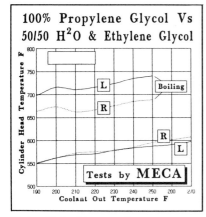

Tests with the Meca cooling system on an aluminum-headed Corvette engine indicated a dramatic reduction of the metal temperature at the critical point between the two exhaust valves on each bank. The top two lines indicate left and right banks of cylinders with a regular coolant. By the time the coolant temperature was up to 250deg., the engine was boiling. On the other hand, using the propylene glycol mixture and the reverse cooling the temperature at the measured point in the cylinder heads had dropped about 150deg. Additionally, each bank of cylinders ran at more even temperatures. The pump needed to accommodate the reverse flow of this system seems to have countered the unequal flow produced by the stock pump.

and refined their systems and as of 1991, almost comparable results have been achieved with conventional flow systems. To date, many difficult cooling cases have been corrected with the Meca system.

Another item is the wetting agent sold by Red Line, the synthetic oil people. This radiator additive is claimed to reduce water temperatures in the order of 10deg. If this claim is realistic then this is certainly a creditable achievement for a simple radiator additive.

Water Pumps

Although there may be variations on the theme, there are three basic forms that a Chevrolet water pump can take. Passenger cars up to 1968–1969, Corvettes up to about 1971, and trucks up to about 1972 all used what is known as the short water pump (a few exceptions exist, such as the V-8 Monza). The water pump style was then changed to what is known as the long water pump.

Basically, the long pump is taller from the face that bolts to the block to the fan mounting face. It also has a different pulley bolt pattern—the short pump had a 1³/₄in bolt pattern, whereas the long pumps utilized a 2¹/₈in bolt pattern. Also, the later

pump employed bigger impeller bearings and shaft. If you intend to purchase a straight replacement pump, you'll find that most usually have a flange with both bolt patterns.

Most stock water pumps utilize a pressed-steel impeller. These impellers appear to suffer from a design problem which cuts their flow capability. Some aftermarket manufacturers, such as the Brass Works and Speed-O-Motive, modify stock pumps by adding a disc to the back of the impeller. They claim that this adds 20-30 percent to the flow capability of the pump. If you want to really upgrade the pump, Bow-Tie high-output pumps are available from Chevrolet, as well as many aftermarket pumps.

If you're going the Bow-Tie route, as of 1991, there are basically three pumps to choose from. The most commonly used pump, part number 14011021, is aluminum bodied and of the short design. In addition there are two other pumps, part number 3998207 for pre 1971 Corvettes and pre 1969 passenger cars, and part number 6258551 for Corvettes from 1971-77. This particular pump features the large bearing of the later-style pump, as does the aluminum-cased pump.

The basic aluminum pump will install in most pre 1969 vehicles, but if

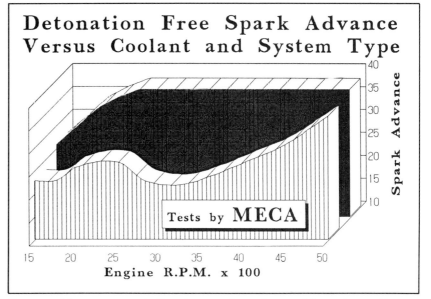

Because of the more even cooling given with propylene glycol and the reverse cooling, this engine could run a virtually optimum advance curve. The front curve shows the advance characteristics that had to be built into the distributor to keep the engine out of detonation at wide open throttle. Because the coolant allowed the more optimum spark in the rear curve, a substantial increase in engine output was realized.

155

it's to be used in a later vehicle, pump spacers, as supplied by Moroso (among others), are needed to make the short outlet to the same length as the long-outlet pump. In addition to the stock and Bow-Tie pumps available from Chevrolet, over a period of time I have used pumps from Howard Stewart, Inc. and Moroso with good effect.

If you're considering a drag-racing only application, then the Moroso pump can be supplied with a cog tooth-belt and an adapter plate to drive it with an electric motor. This boosts horsepower by 12-15hp on an 8000rpm motor. However, flow is limited and so is the block pressure developed.

The significance of reduced flow is easy to understand but the effect of pump-developed pressure is far less obvious. To see why the pressure in a cooling system is important it's necessary to realize that the boiling point of water goes up considerably as pressures rise. At 70psi, the boiling point goes up all the way to 316deg. F.

The primary pressure in a cooling system is developed by the radiator cap. The highest rated cap should be used consistent with the cooling system's ability to take the pressure. Obviously block and heads are unlikely to rupture from pressurization, but radiators are. A radiator that fails at less than 20psi is not up to the job.

On top of the pressure developed in the cooling system is the pump-generated pressure. The basic flow path of the water is in through the block, up through the head, out at the thermostat and into the radiator. The block, heads and thermostat represent a considerable flow restriction, so as the pump speed rises, more block pressure occurs. This block pressure increases the boiling point of the coolant and reduces the likelihood of hot spots flashing to steam.

Having an electric-driven pump will only produce 10-15 gallons per minute flow and pump induced pressures of at best about 2-3psi. Since even a drag-race engine needs to be thoroughly warmed up (especially the oil) it is necessary to have enough cap pressure to keep hot spots from localized boiling as it's these that ultimately cause the onset of detonation.

For a drag-race engine then, you need to use as high a cap pressure as possible, and you need to pressure test the radiator at that pressure, because most pressure caps are on the inlet side of the pump and therefore do not see the pump outlet pressure. This means a 15psi cap can still seal up in a system that develops 60psi in the block. However, if the cap is on the outlet side of the pump it will be exposed to pump pressure and release well before the required pressures are achieved in the block.

By using a high-pressure radiator the effectiveness of the electric pump can be enhanced. The higher pressure doesn't stop the engine showing a high water temperature, but it does prevent boiling, and so resists the effect of hot spots. Going this route makes the use of a smaller, lighter radiator possible, or an increase in compression before detonation is likely to start.

Though pressure sets the upper limit of boiling, it's flow that influences the temperatures seen. The faster water is flowed through the block the more even the temperatures will be. The larger and more efficient the radiator is, the lower the bulk temperature will be. For an effective cooling system then, you have pressure, pump flow, distribution and radiator capability to consider.

I talked about correcting stock pump distribution problems earlier, but the best cure is to fix the pump so it delivers an even flow of water. The latest Howard Stewart pump appears to do this.

If you are buying a pump specifically for a high-performance engine, consider getting one with a cam stop

Water Boiling Point Temperature Vs Pressure

As the pump-developed pressure in the block increases, so does the boiling point. Often if an engine nears the pressure-induced boiling point, I find that lifting off the throttle and dropping the rpm can momentarily allow the coolant to boil because the block pressure has dropped. If this happens, the pressure will almost certainly cause the radiator cap to momentarily vent. This can cause a critical water loss over a relatively short period of time. To avoid this happening on an endurance-race engine, it's necessary to have an expansion tank to recoup the lost water.

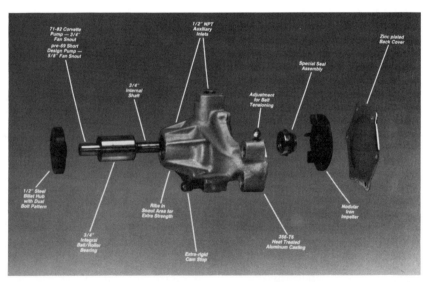

There's more to making a heavy-duty water pump than just simply making it out of aluminum. As you can see from this exploded view of a Howard Stewart pump, many additional features are needed to make it compatible with most racer's requirements.

on it. The cam stop can be used to adjust against the timing case so that it fixes cam walk. Both the Howard Stewart and the Moroso pump feature an anti-cam-walk bolt.

Perhaps the ultimate in water pump setup for the small-block Chevy is the one from Cosworth. This setup features a water pump driven by the same shaft that drives the dry-sump oil pressure and scavenge pumps. This means the water pump is located in a position other than on the front of the engine, as per most conventional pumps. This pump has become popular with IMSA race engine builders and though effective, it is one of the most expensive on the market. In case you skipped over it, details on the oil pump for this type of setup are included in chapter 16 on lubricants.

Pump Drives

No matter which water pump you choose, the speed at which it is driven is going to be relevant to performance. Overspeeding a stock water pump will not necessarily aid cooling—in fact, if overspeeding is excessive, it can do the reverse. If the pump speed goes much over 6000rpm, cavitation can occur and pumping efficiency will drop so much that it will aggravate existing overheating problems. Increasing the system pressure can delay pump cavitation but this is not always a satisfactory solution as it may not cure the problem.

Many of the aftermarket water pumps, plus the Bow-Tie pumps, utilize a different impeller than the stock Chevy one. The stock Chevy impeller is made of pressed steel, but most aftermarket pumps use a cast-iron impeller of a superior vane shape. With these higher efficiency pumps it becomes practical to underdrive the pump and still circulate the required amount of water. By underdriving the pump, rpm is lower than the engine, thus power will be conserved.

Certain pump drive-ratio options are open to you by swapping stock pulleys. Since the most common engine to modify is the 350, and since this was, in almost all cases, fitted with the long pump, a certain amount of interchanging can be done. Early short-leg pumps were driven pretty much 1:1 by the engine crankshaft. The later pumps could be driven anywhere from about 25 percent over to about 35 percent under.

The biggest GM pulley, GM part number 3972180, is a three-groove 8in diameter item. If your particular engine has to drive accessories such as power steering, air conditioning and a water pump, then you don't have much choice. Basically, the principal pulley used with the three-groove pul-

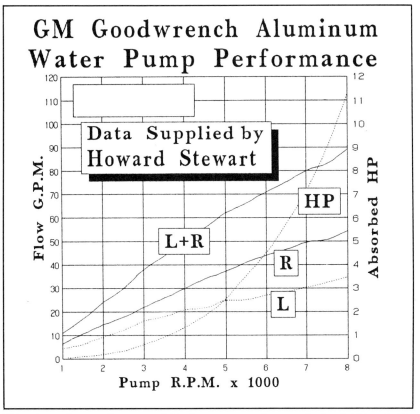

GM Goodwrench Aluminum Water Pump Performance

Data Supplied by Howard Stewart

Flow G.P.M. / Absorbed HP / Pump R.P.M. x 1000

This chart shows why a small-block Chevy tends to run hotter on the left bank than on the right. The prime reason is that more water goes to the right-hand side, thus cooling that side more.

As of 1992 this is the latest Howard Stewart pump. The special diffuser seen here in the middle of the photo helps distribute flow evenly between left and right banks.

Crank to Water Pump Pulley Ratios

Crank Pulley*	Grooves	Diameter (in)	Water Pump Part Number	Diameter (in)	Grooves	Ratio
3956664	1	6⅞	3995631	6³⁄₁₆	1	11% Over
395664	1	6⅞	3989305	7⅛	1	3½% Under
395664	2	6⅞	351680	6³⁄₁₆	2	11% Over
3956666	2	6⅞	3995631	6³⁄₁₆	1	11% Over
3956666	2	6⅞	3989305	7½	1	3½% Under
3956666	2	6⅞	351680	6³⁄₁₆	2	11% Over
3972180	3	8	3995631	6³⁄₁₆	1	29% Over
3972180	3	8	3989305	7⅛	1	12% Over
3972180	3	8	351680	6³⁄₁₆	2	29% Over

*These numbers are stamped on the pulleys, so all the above can be located at the wrecking yard by these numbers.

This chart gives you the interchange numbers for your pulleys and the ratios that they will deliver. By figuring out how much the pump is slowed, and looking at our power absorbtion curve for a stock pump, you should be able to easily figure out how much power is saved.

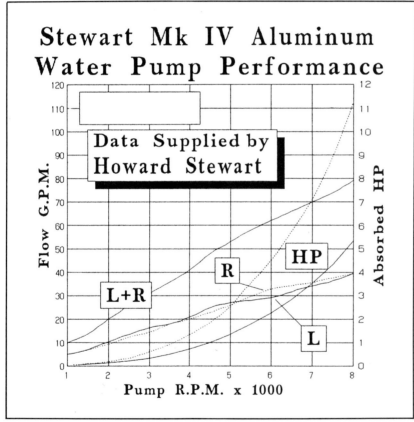

Stewart Mk IV Aluminum Water Pump Performance

Data Supplied by Howard Stewart

Flow G.P.M. / Absorbed HP / Pump R.P.M. x 1000

To get even cooling and temperatures at the heads, which will allow maximum compression and spark advance for all cylinders, it's necessary to employ the use of a pump that gives more even flow on both banks. Here is the output of the latest pump from Howard Stewart as of 1991. Note that each leg of the pump delivers virtually the same amount of flow. In addition, you need to note the amount of horsepower drawn by the Stewart pump. The lower solid horsepower curve is the amount absorbed by the Stewart pump, whereas the dotted line, which goes just over 11hp at 8000rpm, is a typical stock GM pump. By using the Stewart pump the engine should have more even cooling and it should liberate about another 5hp at the flywheel.

ley is the two-groove pulley, part number 351680. This 6³⁄₁₆in diameter pulley overdrives the pump by as much as 25 percent.

The single-groove pulley, part number 3956664, is an older style pulley that can be picked up from the wrecking yard, but cannot be obtained new. It has been unavailable for several years. If this pulley is used with the pump pulley, part number 3989305, then the pump will be underdriven by about 3.5 percent.

If you're going to use an aftermarket pulley, then many options are open to you. I suggest contacting Arias, B&M Automotive Products, Billet Specialties, Moroso, Manley Performance Products, and Jones Machine Racing Products for alternative pulleys to get the drive ratio where you want it. If you decide to use a serpentine belt, as on later cars, then Manley and Dienerbilt are probably the companies to contact.

Sources

ABS Products
PO Box 1984
Southgate, CA 90280

Air Flow Research
10490 Ilex Avenue
Pacoima, CA 91331

Allan Lockheed & Assoc.
PO Box 10828
Golden, CO 80481

APT
561 Iowa Avenue
Annex E
Riverside, CA 92507

Arias Industries
13420 South Normandie Avenue
Gardena, CA 90249

ARP
8565 Canoga Avenue
Canoga Park, CA 91304

B&M Automotive Products
9152 Independence Avenue
Chatsworth, CA 91311

BHJ Products
37530 Enterprise Court
Newark, CA 94560

Bill Miller Engineering
1420-B 240th Street
Harbor City, CA 90710

Callies Performance Products
PO Box 670
Fostoria, OH 44830

Canton Racing Products
9 Tipping Drive
Branford, CT 06405

Childs & Albert
24849 Anza
Valencia, CA 91352

Classic Motorbooks/Motorbooks
   International
PO Box 1
Osceola, WI 54020
800-826-6600

Cloyes Gear & Products
4520 Beidler Road
Willoughby, OH 44094

Cosworth Engineering Inc.
23205 Early Avenue
Torrance, CA 90505

Crane Cams
530 Fentress Boulevard
Daytona Beach, FL 32014

Cunningham Rods
550 West 172nd Street
Gardena, CA 90248

Engine Tech
3656 Hixon Pike
Chattanooga, TN 37415

Faria Engineering
13897 Road 144
Tipton, CA 93272

Federal Mogul Corp.
PO Box 1966
Detroit, MI 48235

Fischer Engineering
9003 Norris Avenue
Sun Valley, CA 91352

Goodrich USA
20309 Gramercy
Torrance, CA 90505

Howard Stewart
108 North Main Street
High Point, NC 27262

Howell Engine Development
5989 Kensington
Detroit, MI 48224

Howell Engineering Co.
PO Box 269
Bryn Mawr, CA 92318

Impulse Engineering
Piston Vice
12731-A Loma Rica Drive
Grass Valley, CA 95945

Jones Racing Products
Rt 611 & Annawanda Road
Ottsville, PA 18942

K&N Engineering
PO Box 1329
Riverside, CA 92502

Manley Performance
13 Race Street
Bloomfield, NJ 07003

Mcca (Cooling Division)
255 Route 41 North
Sharon, CT 06069

Melling Performance
2712 North 58th Street
Tampa, FL 33619

Metal Improvement Co. Inc.
10 Forest Avenue
Paramus, NJ 07652

Moroso Performance Products
80 Carter Drive
Guilford, CT 06437

Nitron
330 Canal Street
Lawrence, MA 01840

Oberg Enterprises
12414 Highway 99 South
Bay 80
Everett, WA 98204

Polymer Dynamics
4116 Siegel
Houston, TX 77009

PRD
7630 Miramar Road #2500
San Diego, CA 92126

Pro Cam
17765 148th Avenue
Spring Lake, MI 49456

RHS Performance Marketing
3410 Democrat
Memphis, TN 38118

Ross Racing Pistons
11927 South Prairie Avenue
Hawthorne, CA 90250

Sealed Power Corp.
100 Terrace Plaza
Muskegon, MI 49443

Snap On Tools
PO Box 26922
Indianapolis, IN 46226

Speed-O-Motive
12061 East Slauson
Santa Fe Springs, CA 90670

Steve Jennings Equipment
1401 East Bochard Street
Santa Ana, CA 92705

Summers Brothers
530 South Mountain Avenue
Ontario, CA 91762

Sunnen Products Co.
7910 Manchester
St. Louis, MO 63143

Superflow Corp.
3512 North Tejon
Colorado Springs, CO 80907

Swain Tech Coatings
35 Main Street
Scottsville, NY 14546

System One Filtration
1822 East Main Street, #A
Visalia, CA 93291

TRW Automotive Aftermarket Division
8001 East Pleasant Valley Road
Cleveland, OH 44131

Vandervell America
2488 Tuckerstone Parkway
Tucker, GA 30084

Vibra Tech/Fluidamper
537 East Delavan Avenue
Buffalo, NY 14211

Wiseco Piston Inc.
7201 Industrial Park Boulevard
Mentor, OH 44060

Index